Kazimierz J. Ducki

Fatigue behaviour and creep resistance of Fe-Ni superalloy

AF191365

Kazimierz J. Ducki

Fatigue behaviour and creep resistance of Fe-Ni superalloy

LAP LAMBERT Academic Publishing

Publisher:
LAP LAMBERT Academic Publishing
is a trademark of
Dodo Books Indian Ocean Ltd. and OmniScriptum S.R.L publishing group

120 High Road, East Finchley, London, N2 9ED, United Kingdom
Str. Armeneasca 28/1, office 1, Chisinau MD-2012, Republic of Moldova, Europe
Managing Directors: Ieva Konstantinova, Victoria Ursu
info@omniscriptum.com

Printed at: see last page
ISBN: 978-3-659-19549-5

Table of Contents

1. Introduction 3

2. General characteristics of Fe-Ni alloys...6

   2.1. Fe-Ni alloys of solid-solution strengthened 6

   2.2. Fe-Ni alloys of precipitation strengthened by γ' phase......................7

   2.3. Fe-Ni alloys with a low thermal expansion...9

   2.4. Fe-Ni and Ni-Fe alloys of precipitation strengthened by γ'

      and/or γ'' phases 10

   2.5. Fe-Ni and Ni-Fe alloys of precipitation strengthened by

      carbides...12

   2.6. Fe-Cr alloys of oxide dispersion strengthened (ODS).....................12

   2.7. Summary of characteristics for Fe-Ni alloys 14

3. Heat treatment of Fe-Ni superalloys 16

5. Microstructure and mechanical properties of Fe-Ni superalloy................25

6. Fatigue behaviour of Fe-Ni superalloy 30

7. Creep resistance of Fe-Ni superalloy..42

8. Summary 53

9. References 57

1. Introduction

Multicomponent creep-resistant Fe-Ni-Cr alloys strengthened by the precipitation of intermetallic phases and often defined as Fe-Ni superalloys are an important group of construction materials. They are characterised by a number of specific properties [1÷4]: good mechanical properties at room and elevated temperatures, good creep resistance and heat resistance, satisfactory resistance to corrosion, and high ductility and impact strength at low temperatures. Fe-Ni superalloys can be used in a temperature range from liquid helium (-269°C) to 540÷815°C [4÷7]. At such temperatures, the superalloys can be used in conventional and nuclear power generation [8÷19], aviation technology [2,4,10,11], the chemical and petrochemical industries [20÷22], the electromachinery industry [11,23], cryogenics [5÷7], and for the manufacture of tools for the processing of non-ferrous metals and alloys [24÷26,31].

Fe-Ni alloys are widely applied for operation at elevated and high temperatures, where they can be used as a structural material intermediate between types of martensitic steel used for operation at temperatures up to 600°C, and types of creep-resistant nickel superalloys intended for operation at temperatures above 700°C [1,16,18]. Creep-resistant Fe-Ni alloys precipitation strengthened, compared to Ni- and Co-based alloys, have lower resistance to oxidation and gas corrosion, but have a high creep resistance at an intermediate range of temperatures, i.e. 540÷700°C. It may be expected that along with the continuous intensification of technological processes, increases in the operating parameters of machines in the energy generation industry, and the development of new technologies in the chemical, petrochemical and processing industries, this group of materials will gain a more important role. The analysis of the current state of research on Fe-Ni alloys precipitation strengthened demonstrated that they are an interesting scientific problem, but also have utilitarian aspects, as their cost of

3

manufacture is considerably lower in comparison to nickel and cobalt superalloys [1,2,12,27,28,31].

Alloys strengthened by precipitated ordered intermetallic phase γ' - $Ni_3(Al,Ti)$ with a regular structure (FCC) are the main group of creep-resistant Fe-Ni superalloys. The γ' phase in Fe-Ni alloys precipitates in the form of spheroid, highly dispersed particles coherent with the γ solid solution. In many reports [1,2,27,29÷42] the γ' phase was considered to be the major phase strengthening iron-based alloys. In high-temperature applications, particles of the γ' phase strengthen Fe-Ni alloys up to 750°C. At higher temperatures the γ' phase transforms into the η phase (Ni_3Ti), of a hexagonal structure and lamellar morphology. In Fe-Ni alloys the strengthening effect caused by the precipitation of the η phase is less pronounced in comparison to strengthening by the γ' phase. Reduction in the nickel content in the matrix caused by the precipitation of γ' and η phases, at the simultaneous presence of molybdenum or wolfram as additives strengthening the solid solution, creates conditions for the precipitation of particles from TCP phases – σ, Laves, G, χ and μ, decreasing the plasticity of alloy. At the same time, the precipitation of carbides (MC, $M_{23}C_6$ and M_6C) and borides or boron carbides of different morphology can occur. The structural stability of Fe-Ni alloys improves as a result of the partial replacement of iron by nickel. This process allows for the increase in the concentration of components strengthening the solid solution γ and improves creep resistance, without the formation of undesirable TCP-phases [8,32,39÷42].

Previous studies [27,29,31-33,39÷42] on Fe-Ni alloys were mainly focused on the analysis of the precipitation of intermetallic phases (γ', η and G) and carbides ($M_{23}C_6$, M_6C and MC) during heat treatment or the operation of products, and their effects on strength and plastic properties. However, there has been no comprehensive analysis of the effect of different variants of initial heat treatment on the microstructure and characteristics of Fe-Ni alloys at low-cycle fatigue tests, and creep at elevated temperatures.

In the present work, an investigation was initiated concerning the effect of initial heat treatment on the microstructure, mechanical properties, fatigue life and creep resistance of an Fe-Ni superalloy precipitation-strengthened with an intermetallic phase of the γ' type. Specimens of Fe-Ni alloy were subjected to tests after two variants of heat treatment, i.e. solution heat treatment followed by typical single-stage ageing, and solution heat treatment followed by novel two-stage ageing. It is assumed that the obtained test results of fatigue life and creep resistance may be used to optimise heat treatment and forecast the operating conditions of products made of Fe-Ni superalloy at elevated temperatures.

2. General characteristics of Fe-Ni alloys

Developmental work on creep-resistant wrought Fe-Ni alloys has been carried out in parallel with work on nickel superalloys. There are significant differences in physical, chemical and mechanical properties between iron and nickel superalloys, caused mainly by differences in their chemical compositions. Nickel superalloys were created and developed through the chemical modification of nickel-chromium alloy (NiCr20), while Fe-Ni alloys were developed by modifying the chemical and phase composition of austenitic steel 18-8. Currently, many researchers [1,2,4,12,28,43] claim that most creep-resistant Fe-Ni superalloys contain min. 36% Fe, max. 45% Ni and min. 12% Cr. However, the above contents of elements are not strictly followed. Recent research has led to the development of more than twenty types of iron alloys with complicated chemical compositions. Their original classification, based on the main mechanism of strengthening, chemical and phase composition and development of Fe-Ni alloys is proposed in Table 1.

2.1. Fe-Ni alloys of solid-solution strengthened

The first generation Fe-Ni superalloys are the group of solid-solution strengthened alloys. They are characterised by increased content of nickel, chromium and carbon. This group of materials includes 19-9DL, 17-14CuMo and 16-25-6 alloys most similar to 18-8 steel, which were used for the manufacture of elements in aircraft and industrial gas turbines operating at temperatures up to 650°C [2,4,6]. Components strengthening the solid solution in these Fe-Ni alloys include [1,2,4,29]: Mo (max. 6%), W (max. 1.2%), Nb (max. 0.4%) and Ti (max. 0.3%). Molybdenum and niobium are substituted by wolfram or tantalum, respectively. According to studies by Ulianin et al. [20], Larkin et al. [43] and Watanabe and Kuno [44], Fe-Ni solid-solution strengthened alloys containing wolfram instead of molybdenum have a higher creep resistance and plasticity limit in comparison to alloys containing molybdenum. All these elements cause difficulties in plastic forming at high temperatures. In addition, they stimulate the formation of ferrite δ, which was the reason for reducing the content of ferrite-promoting elements and increasing the nickel content. This led to the development of the second subgroup of Fe-Ni solid-solution strengthened alloys, offering good mechanical properties at high temperatures and high creep resistance (Incoloy 800, 801, 802 and Carpenter 20Cb-3), Table 1. These alloys can operate at temperatures up to 1100°C, and today they are the basic group of metallic materials used for the construction of highly reliable units in chemical, powergeneration and other systems [2,4,45].

2.2. Fe-Ni alloys of precipitation strengthened by γ' phase

Strengthening of an alloy at elevated working temperatures through the introduction of additives to the solid solution is insufficient. A more efficient technique is based on strengthening by particles from intermetallic phases, carbides or nitrides. For this reason the second generation of Fe-Ni alloys strengthened by precipitated particles of the γ' - $Ni_3(Al,Ti)$ phase was developed, e.g. W-545, A-286, Discaloy, Tinidur, V-57 (Table 1). These alloys contain ca. 14÷16% Cr, improving the resistance of the material to oxidation and gas corrosion, and ca. 24÷27% Ni to stabilize the microstructure of the γ matrix. In addition, the second generation superalloys contain the addition of 1 to 3% Mo and V, strengthening the solid solution, and 2.0 to 3.5% Ti and Al, which enable strengthening by precipitated particles of the γ' phase. Most types of these superalloys contain more titanium than aluminium, because the higher titanium content ensures higher strength at high temperatures [4,11,29]. In order to achieve the maximum parameters of these alloys, not only should the content of titanium and aluminium be controlled, but also the proportion of the contents and their total content.

Table 1. Nominal chemical composition, creep limit and creep rupture strength for selected wrought Fe-Ni, Ni-Fe and Fe-Cr superalloys [1,2,4,32,45,48,54,56,57,60].

| No. | Alloy | Nominal composition [wt. %] | | | | | | | | | | | | | | $R_{t/10^4}/R_{u/10^3}$ h [MPa] at temperature [°C] | | |
		C	Mn	Si	Cr	Ni	Fe	Co	Mo	W	Nb	Ti	Al	B	Other	650	730	815
						Group I. Fe-Ni alloys of solid-solution strengthened												
1.	19-9DL	0.30	1.1	0.6	19.0	9.0	66.8	–	1.25	1.25	0.4	0.3	–	–	–	303/255	155/117	89/59
2.	17-14CuMo	0.12	0.75	0.5	16.0	14.0	62.4	–	2.5	–	0.4	0.3	–	–	3.0Cu	306/245	163/120	91/61
3.	16-25-6	0.06	1.35	0.7	16.0	25.0	50.7	–	6.0	–	–	–	–	–	0.15N	310/234	172/117	93/62
4.	Incoloy 800	0.05	0.8	0.5	21.0	32.5	45.7	–	–	–	–	0.4	0.4	–	0.4Cu	220/158	89/67*	67/48
5.	Incoloy 801	0.05	0.8	0.5	20.5	32.0	46.3	–	–	–	–	1.1	–	–	0.2Cu	276/191	138/94	69/46
6.	Incoloy 802	0.35	0.8	0.4	21.0	32.5	44.8	–	–	–	–	0.8	0.6	–	0.4Cu	248/172	144/110*	103/69**
7.	Carpenter 20Cb-3	0.07	0.8	0.4	20.0	34.0	42.4	–	2.5	–	1.0	–	–	–	3.5Cu	282/195	158/121*	112/75**
						Group II. Fe-Ni alloys of precipitation strengthened by γ' phase												
8.	W-545	0.08	1.5	0.4	13.5	26.0	55.8	–	1.5	–	–	2.85	0.2	0.080	–	552/448	338/255	–
9.	A-286	0.05	1.4	0.4	15.0	26.0	55.2	–	1.25	–	–	2.0	0.2	0.003	0.3V	420/317	241/144	89/55
10.	Discaloy	0.04	0.9	0.8	13.5	26.0	55.0	–	2.75	–	–	1.75	0.25	–	–	359/283	207/138	103/–
11.	Tinidur	0.04	1.0	0.75	14.5	26.0	54.0	–	1.25	–	–	2.15	0.2	0.003	0.03V	430/321	260/155	–
12.	V-57	0.08	0.35	0.75	14.8	27.0	48.6	–	1.25	–	–	3.0	0.25	0.010	0.5V	586/483	345/200	–
						Group III. Fe-Ni alloys with low thermal expansion												
13.	Incoloy 903	0.04	–	–	0.1	38.0	41.0	15.0	0.1	–	3.0	1.4	0.7	–	–	627/510	–	–
14.	Incoloy 907	0.01	0.3	0.15	0.1	38.0	42.0	13.0	–	–	4.7	1.5	0.03	–	–	688/560	–	–
15.	Incoloy 909	0.01	–	0.4	0.1	38.0	42.0	13.0	–	–	4.7	1.5	–	0.001	–	424/345	–	–
16.	Pyromet CTX-1	0.03	–	–	0.1	37.0	39.0	16.0	0.1	–	3.0	1.7	1.0	0.003	–	616/505	–	–
17.	Pyromet CTX-3	0.05	–	0.15	0.2	38.0	41.0	13.5	–	–	4.9	1.6	0.1	0.007	–	695/570	–	–
						Group IV. Fe-Ni and Ni-Fe alloys of precipitation strengthened by γ' and/or γ'' phases												
18.	CG27	0.05	0.1	0.1	13.0	38.0	38.4	–	5.7	–	0.7	2.5	1.6	0.010	–	676/531	434/303	241/151
19.	Incoloy 901	0.05	0.1	0.1	12.5	42.5	36.0	–	5.7	–	–	2.8	0.2	0.015	–	552/441	338/213	131/75
20.	Pyromet 860	0.05	0.05	0.05	12.6	43.0	30.0	4.0	6.0	–	–	3.0	1.25	0.010	–	655/559	414/310	227/117
21.	D-979	0.05	0.3	0.2	15.0	45.0	27.0	–	4.0	4.0	–	3.0	1.0	0.010	–	648/524	414/303	227/151
22.	Incoloy 706	0.03	0.2	0.2	16.0	41.5	40.0	–	–	–	2.9	1.8	0.2	–	–	691/580	305/181*	227/151
23.	Inconel 718	0.04	0.2	0.2	19.0	52.5	18.5	–	3.0	–	5.1	0.9	0.5	–	–	702/595	335/195*	–
						Group V. Fe-Ni and Ni-Fe alloys of precipitation strengthened by carbides												
24.	N-155	0.10	1.5	0.5	21.0	20.0	32.2	20.0	3.0	2.5	1.0	–	–	–	0.15N; ≤0.5Cu	359/296	193/151	138/110
25.	Haynes 556	0.10	1.5	0.4	22.0	20.0	29.0	20.0	3.0	2.5	0.1	–	0.3	–	0.5Ta; 0.02La; 0.002Zr	340/275	162/125*	72/55**
26.	Hastelloy X	0.10	0.5	0.5	22.0	47.0	18.5	1.5	9.0	0.6	–	–	–	–	–	320/222	130/90*	96/69**
						Group VI. Fe-Cr alloys of oxide dispersion strengthened (ODS)												
27.	Incoloy MA956	0.05			20.0		74.4					0.5	4.5		0.5Y₂O₃	160/110		94/65***
28.	Incoloy MA957	0.05			14.0		84.4	0.3				1.0			0.25Y₂O₃	150/100		86/60***
29.	PM2000 Alloy	0.05			20.0		73.4					0.5	5.5		0.5Y₂O₃		145/100**	115/80***

* at temperature 760°C; ** at temperature 870°C; *** at temperature 980°C.

8

According to Ohta et al. [46] the correct Ti/Al proportion ensuring good creep resistance of Fe-14Cr-30Ni alloys strengthened by the phase γ' precipitates is from 2 to 5. Types of these alloys are intended for operation in the temperature range 540 to 730°C [4,11,17]. They are mainly used for highly loaded rotors of industrial gas turbines and aircraft engines. According to Sims et al. [1] most turbine elements are made of A-286 alloy. For example, SNECMA [17] uses this type of alloy to manufacture 500÷1500 kg discs for industrial turbines, while Kobe Steel Ltd. [9] manufactures discs for gas turbines size Ø 1540 × 250 mm and weight about 3.5 Mg. Another Japanese company, Mitsubishi Heavy Industries Ltd. [15], has used the modified A-286 superalloy to manufacture the largest forging in the world of a steam turbine rotor, weighing 12 Mg. The rotor was installed in the Wakamatsu II steam turbine, operating under ultra-supercritical conditions (USC). The rotor operated successfully at steam parameters 650°C/34.5 MPa, which was a global achievement in 1990/91. Currently, further work on this group of Fe-Ni alloys is focused on the modification of their chemical composition by adding 0.5 to 1% of niobium [22,47÷50]. The purpose of this is to increase the relative volume of the γ' phase and its thermal stability, in order to delay the ageing process and improve the weldability of the alloy. In the γ' phase niobium atoms partially replace Al and Ti atoms, reducing the susceptibility of welded joints to cracking during annealing caused by the decrease in coherence stress associated with the precipitation of the γ' phase. At elevated temperatures Nb-A286 alloy modified with niobium has a higher tensile strength and resistance to creep and low cycle fatigue when compared to the conventional A-286 alloy [47].

2.3. Fe-Ni alloys with a low thermal expansion

This is the third generation of Fe-Ni superalloys characterised by increased content of cobalt. The main types of these alloys are Inconel 903, 907 and 909, and Pyromet CTX-1 and CTX-3 (Table 1). The low values of the coefficients of linear thermal expansion in this group of alloys were obtained by a reduction in the content of ferrite-promoting elements, mainly chromium and molybdenum. These alloys contain ca. 37÷38% Ni and ca. 13÷16% Co, to stabilize the solid solution γ, and the addition of Nb (3÷5%), Ti (1.4÷1.7%) and sometimes Al (0.03÷1.0%) – forming the precipitates of the γ' phase - $Ni_3(Ti,Nb)$ and strengthening the matrix [48]. The alloys maintain the low values of the coefficient of linear thermal expansion at temperatures from - 40°C to 680°C and have high tensile strength at temperatures up to 650°C [2,4]. Lack of chromium makes them susceptible to oxidation at temperatures above 540°C [4], and prone to stress accelerated grain boundary oxidation (SAGBO) [48]. Increased creep resistance of the Inconel 909 alloy was obtained by reducing the content of Al, increasing the content of Nb and Si, and the use of the alloy in the recrystallised form [48].

2.4. Fe-Ni and Ni-Fe alloys of precipitation strengthened by γ' and/or γ'' phases

The fourth generation of Fe-Ni superalloys include alloys with a high nickel content (over 38%), precipitation-strengthened by particles of the γ' - $Ni_3(Al,Ti)$ and/or γ'' - Ni_3Nb phases. A subgroup of Fe-Ni alloys with a high nickel content precipitation-strengthened by the γ' phase is represented by CG27, Incoloy 901, Pyromet 860 and D-979 (Table 1). These alloys contain Ni (ca. 38÷45%) stabilizing the γ matrix, Cr (ca. 12÷15%) increasing creep resistance, Mo (ca. 4÷6%) and Co (max. 4%) strengthening the solid solution, as well as additives (ca. 3% Ti and ca. 1% Al) forming the γ' phase. Alloys from this subgroup have a higher relative volume of the γ' phase. Through this, they have greater strength parameters at elevated temperatures and can operate at up to 815°C [1,4]. Their primary applications include discs of industrial gas turbines and aircraft engines. The second subgroup of superalloys with a high content of nickel includes Incoloy 706 and Inconel 718

(Table 1), precipitation-strengthened mainly with particles of the γ'' phase. Phase γ'' - Ni_3Nb, with a tetragonal structure and a spatially centred network, is not found in nickel-based alloys. According to the Engel-Brewer correlation [1,51], the γ'' phase is formed in creep-resistant alloys only when a high iron content is present. Usually, Ni-Fe alloys strengthened by the precipitation of the γ'' phase may contain Ni (ca. 42÷52%), Cr (16÷19%), Fe (18÷40%) and Mo (max. 3%) and the addition of Nb (3÷5.5%), Ti (0.5÷2%) and Al (0.2÷0.8%). The high niobium content results in the formation of the main strengthening phase γ''. Titanium and aluminium in lower contents form the γ' phase. Phase γ'' coherent with the matrix causes powerful strengthening of Fe-Ni alloys, but only at temperatures up to 650°C [1,2,28]. In Ni-Fe alloys exposed to long-term work at temperatures over 650°C, the γ'' phase transforms into the stable γ' phase - $Ni_3(Al,Ti,Nb)$ or the equilibrium phase δ - Ni_3Nb of an orthorhombic structure. The higher content of Al and Nb, and the value of the Al/Ti ratio in the Inconel 718 alloy improves mechanical properties by increasing the volume of the stable phase γ' and decreasing the volume of undesirable phase δ [2]. Through this chemical modification and additional heat and plastic treatment, the working temperature of Inconel 718 alloy can be increased up to 760°C [4,12]. Further improvement in the creep resistance and weldability of Inconel 718 alloy can be obtained through the replacement of niobium with tantalum [48]. Fe-Ni and Ni-Fe alloys precipitation-strengthened by phases γ'' and γ' have a good fatigue resistance at elevated temperatures. They are widely used for the parts of gas turbines and aircraft engines (discs, shafts, supports, and other components) [48]. For example, in the USA Incoloy 706 was successfully used to manufacture a gas turbine disc, size Ø 2300 × 400 mm and 12 Mg weight, while Inconel 718 was used for making slightly smaller discs (Ø 2000 × 250 mm) weighing 9 Mg [17]. Inconel 718 is the most popular alloy from this group, and in modern times has become the most dominant alloy in production among all types of Ni-based and Fe-Ni superalloys [52]. Loria [53] estimated that currently

Inconel 718 accounts for ca. 45% of wrought nickel-based alloy production and ca. 25% of cast nickel-based products.

2.5. Fe-Ni and Ni-Fe alloys of precipitation strengthened by carbides

The fifth generation of wrought Fe-Ni superalloys include multicomponent Fe-Cr-Ni-Co solid-solution strengthened alloys and some alloys strengthened by precipitated carbides or carbonitrides. The improvement in mechanical properties in this group of alloys was achieved mainly through the partial replacement of iron with cobalt (ca. 20%), introduction of a high molybdenum content (ca. 9%), and an increase in nickel content to max. 47%. The introduction of cobalt reduced the risk of TCP formation, particularly the brittle phase σ (FeCr), and at the same time improved creep resistance. Typical Fe-Ni superalloys from this group include N-155, Haynes 556 and Hasteloy X (Table 1). These alloys are characterised by different chemical composition and may contain Cr (ca. 21÷22%), Ni (ca. 20% or 47%), Fe (ca. 29÷32% or 18.5%), Co (ca. 20% or 1.5%), Mo (ca. 3% or 9%), W (ca. 2.5% or 0.6%) and increased carbon content (0.10%). The creep resistance of these alloys depends mainly on the type and thermal stability of carbides and carbonitrides. Alloys strengthened by carbides and carbonitrides are characterised by good resistance to high-temperature oxidation, and are used at temperatures up to 815°C [1,2]. For high-temperature low-stress applications they are used at much higher temperatures of 1050÷1100°C [1,2]. They were primarily designed for lining combustion chambers in jet aircraft engines.

2.6. Fe-Cr alloys of oxide dispersion strengthened (ODS)

The sixth, and most recent generation of iron-based wrought superalloys are ferritic oxide (mainly Y_2O_3) dispersion-strengthened alloys (ODS) [2,54÷60]. ODS ferritic superalloys are manufactured through mechanical alloying, which ensures the formation of micrometric grains with particles of Y_2O_3 20÷50 nm in diameter [57,60]. The main types of this alloy subgroup include Incoloy MA 956, Incoloy MA 957 and PM2000 Alloy (Table 1). They contain Fe (ca. 75%), Cr (ca. 14% or 20%), Al (max. 4.5%), Ti (ca. 0.5÷1%), C (ca. 0.05%) and ca. 0.5% or 0.25% of Y_2O_3. Because of the high melting point, thermal stability of the microstructure and resistance to corrosion, these alloys can work at temperatures up to 1100÷1200°C [2,54,57,60]. The creep resistance of ODS alloys is even superior to the conventional nickel and cobalt superalloys (Fig. 1). They are primarily designed for applications in the elements of gas turbines working under high-load conditions, combustion chambers of aircraft engines and elements of Diesel engines. Ferritic ODS superalloys are also used in selected elements of accessories for high-temperature sintering and heat-treatment furnaces, and elements of equipment for the chemical and glass industries. Incoloy MA 957 was also successfully used as nuclear fuel cladding material in a fast breeder reactor [60].

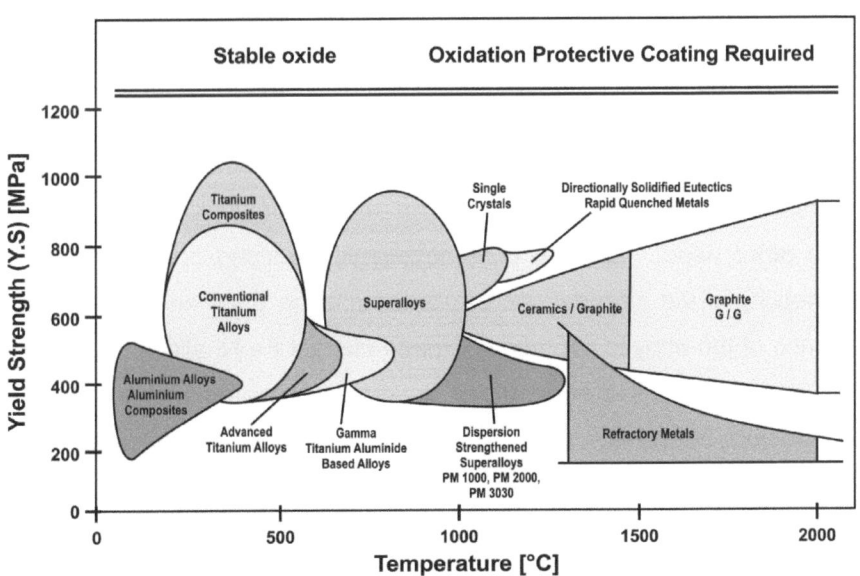

Fig. 1. Range of application of materials depend to working temperature [57]

2.7. Summary of characteristics for Fe-Ni alloys

Analysis of the current state of research demonstrated that the group of creep-resistant wrought Fe-Ni superalloys includes a large number of materials diversified in terms of chemical composition, strengthening mechanism, obtained properties and applications. The primary group of Fe-Ni superalloys includes materials strengthened by the precipitation of ordered intermetallic phases γ' - $Ni_3(Al,Ti)$ and/or γ'' - Ni_3Nb intended for work at temperatures from 540 to 815°C. At temperature above 815°C, the precipitates of intermetallic phases γ' and γ'' become unstable, and age into the equilibrium phases η - Ni_3Ti and δ - Ni_3Nb, which is associated with a significant decrease in their creep resistance. For this reason, at the upper range of working temperatures (900 to 1100°C) Fe-Ni solid-solution strengthened alloys or ODS superalloys are mainly used. Both types of alloys can be hardened to achieve the parameters of precipitation- strengthened alloys. Values of the creep limit ($R_{1/100}$) and 1000-h stress-rupture ($R_{u/1000}$) for selected wrought Fe-Ni alloys are presented in Table 1 and Fig. 2. The creep resistance of Fe-Ni superalloys mainly depends on the concentration of nickel and iron, and the content of elements that produce solid-solution or precipitation hardening of the alloy. The increased content of nickel in Fe-Ni alloys is usually associated with good creep resistance, higher working temperatures, better thermal stability of microstructure, and higher price. On the other hand, high iron content, despite reduced cost and improved machinability of the material, increases the melting point and decreases the resistance of the alloy to oxidation. Creep-resistant Fe-Ni alloys, compared to nickel- and cobalt-based superalloys, have lower resistance to oxidation and gas corrosion, but have a high creep resistance at an intermediate range of temperatures, i.e. 540÷750°C [1÷4,11,28].

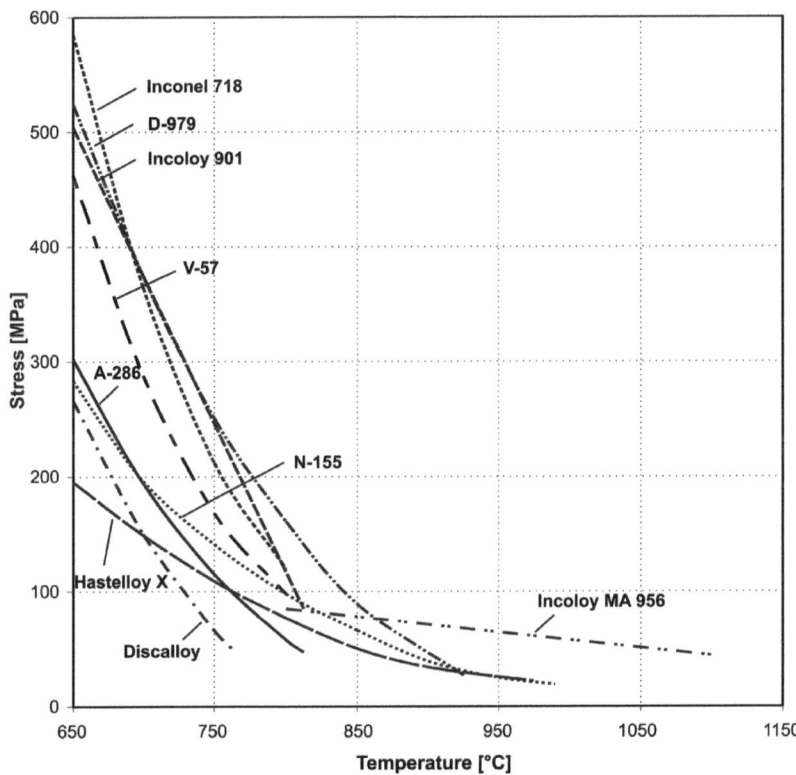

Fig. 2. 1000-h stress-rupture curves of wrought superalloys of Fe-Ni, Ni-Fe and Fe-Cr type [2]

Currently, novel types of ferritic ODS superalloys are strong competition for nickel and cobalt superalloys. They are popular due to the significantly lower price and good creep resistance (up to 1200°C). Fe-Ni superalloys, because of the similar phase composition, are frequently classified to the same group of materials as nickel superalloys [1,2,12,28]. Therefore, their further development is closely connected with the development of wrought creep-resistant nickel superalloys.

3. Heat treatment of Fe-Ni superalloys

The application properties of products made of Fe-Ni, Ni-Fe and Ni-base superalloys largely depend on the correct choice and execution of individual heat treatment operations. All creep-resistant superalloys strengthened by the precipitation of intermetallic phase type γ' - $Ni_3(Al,Ti)$ and/or γ'' - Ni_3Nb, depending on the volume fraction of the γ' and γ'' phases, are subject to two- or multi-stage heat treatment (Fig. 3).

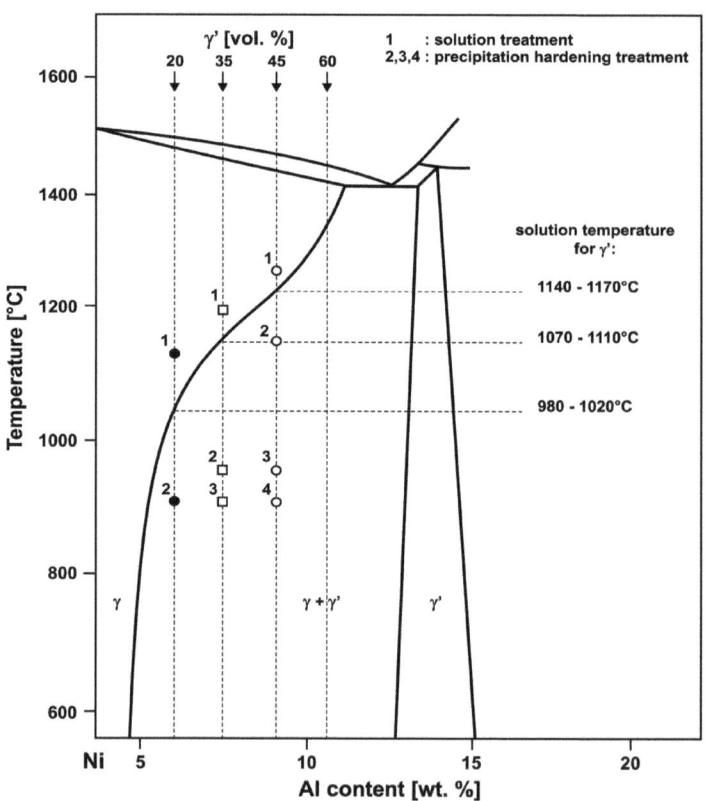

Fig. 3. Scheme of heat treatment for superalloys of precipitation strengthened by γ' phase [40]

In practice, to obtain the optimum microstructure for individual applications, the final heat treatment of creep-resistant superalloys usually consists of two major operations [40,45,61]:

- **solution heat treatment**: at temperatures from ca. 950°C to under 1200°C, depending on the fraction of the γ' phase. The purpose of this operation is to introduce precipitates of the γ' phase created during the hot plastic forming into a solid solution, to dissolve $M_{23}C_6$ carbides, to achieve recrystallised solid solution γ, and to control the size of matrix grains;
- **ageing**: at temperatures in the heterogeneous area where the γ/γ' phases exist, with the aim of controlling the phase γ' precipitates in terms of their fraction, size, morphology and distribution. These structural parameters of the γ' phase depend on the ageing temperature, hold time, cooling rate and the number of heat treatment cycles.

For Fe-Ni alloys containing about 20% of the γ' phase (e.g. A-286, V-57) heat treatment usually consists of two operations, i.e. solution heat treatment and ageing. Ni-Fe alloys that contain about 30% of the γ' phase (e.g. D-979, Inconel 718) are usually subject to three-stage heat treatment, i.e. solution heat treatment and two-stage aging. For nickel-base superalloys that contain about 40÷45% of the γ' phase volume (e.g. Udimet 710, Udimet 720) four-stage heat treatment is usually applied, including solution heat treatment and three-stage ageing [1,2,40,45,61].

Creep-resistant Fe-Ni superalloys typically contain not more than 10÷20% of the γ' phase. For the A-286 superalloy the following heat treatment should be applied (Fig. 4):

- solution heat treatment – from temperatures above the areas where η (Ni_3Ti) and G ($Ni_{16}Ti_6Si_7$) phases exist;
- ageing – within the temperature range where the γ' phase exists, and avoiding the formation of the η phase.

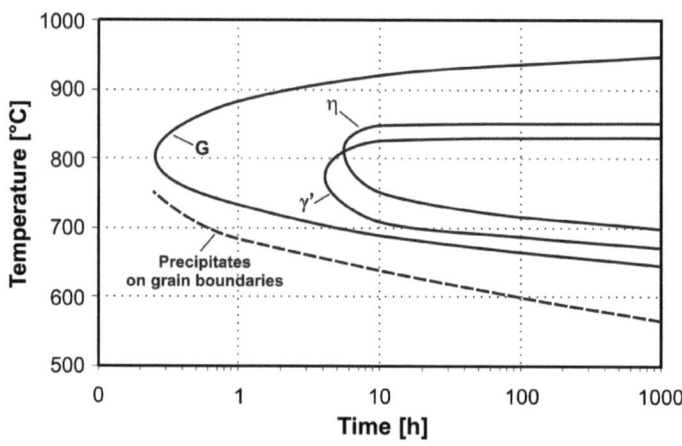

Fig. 4. The TTP diagram of precipitation of secondary phases during annealing A-286 alloy in isothermic condition [61]

The A-286 superalloy is usually subject to solution heat treatment at 980°C and cooling in oil. The use of higher temperatures for alloy solutioning is not recommended due to the growth of austenitic grains and a decrease in the strength parameters during the further ageing stage [39]. The typical and most popular ageing procedure for A-286 superalloy involves annealing at 710÷730°C for 16 h, followed by cooling in the air or furnace down to 600°C, and further air cooling. This heat treatment process ensures the uniform distribution of dispersed particles of the γ' phase with sparsely substituted boundaries of austenite grains within the γ matrix. After this heat treatment, optimum mechanical properties of the alloy are achieved, both at room and elevated temperatures [1,2,39,40,45,61].

For certain creep-resistant superalloys of the Fe-Ni type (e.g. Discaloy, Incoloy 909) and Ni-Fe (e.g. Inconel 718) strengthened by precipitates from the γ' and/or γ'' (Ni$_3$Nb) phases, two-stage ageing is applied that involves the complete and controlled cooling cycle between two isothermal points. The purpose of heat treatment is to achieve the optimum size and distribution of particles from the γ' and/or γ'' phases, ensuring maximum hardness. One example of this is the ageing of the Inconel 718 alloy, which is annealed at 720°C for 8 h, than cooled in the furnace at 650°C for 8 h hold time, followed by air cooling [1,2,39,40,45]. This type of heat treatment ensures maximum strength combined with good alloy plasticity under creep conditions at 650÷705°C.

4. Material and procedure

The examinations were performed on rolled bars, 16 mm in diameter, of an austenitic Fe-Ni superalloy of A-286 type. The chemical composition of the material is given in Table 2.

Table 2. Chemical composition of the investigated Fe-Ni austenitic alloy.

Content of an element [wt. %]															
C	Si	Mn	P	S	Cr	Ni	Mo	V	W	Ti	Al	Co	B	N	Fe
0.05	0.55	1.25	0.026	0.016	14.3	24.5	1.34	0.41	0.10	1.88	0.16	0.08	0.007	0.0062	55.32

Specimens of Fe-Ni alloy were subjected to tests after two variants of heat treatment, i.e. solution heat treatment and single-stage ageing (variant A) and solution heat treatment followed by two-stage ageing (variant B). Parameters of heat treatment for the investigated Fe-Ni alloy were determined based on the previously carried out studies [31,33÷39] and data from professional literature [1,2,40,45,61]. For the investigated alloy after solution heat treatment in the conditions: 980°C/2 h/water, two variants of specimens ageing were used for comparison, i.e.:

- single-stage ageing (variant A): 715°C/16 h/air;
- two-stage ageing (variant B): 720°C/8 h + cooling in the furnace up to a temperature of 650°C + 650°C/8 h/air.

A schematic course of heat treatment of specimens made of the investigated alloy is presented in Fig. 5.

A static tensile test at room temperature and an elevated temperature (600÷750°C) were carried out using a strength testing machine MTS-810. Cylindrical five-time specimens with a diameter d_0 = 10 mm and measuring length l_0 = 50 mm were used for the tests (Fig. 6). A yield strength (Y.S), tensile strength (T.S), unit elongation (EL.) and reduction of area (R.A) were determined.

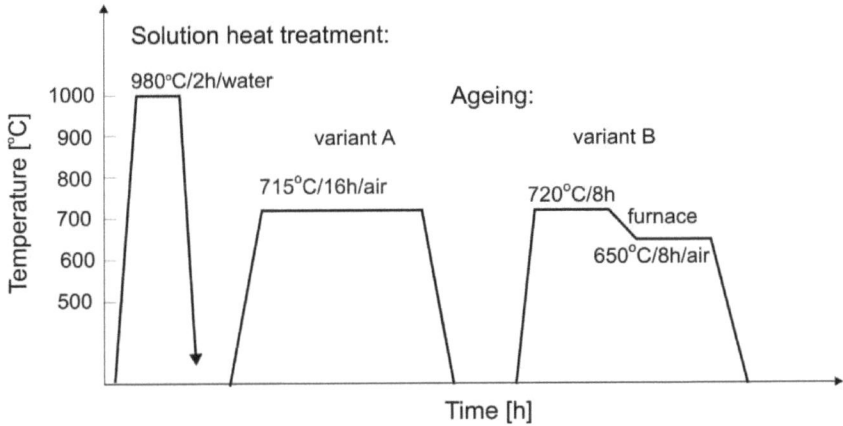

Fig. 5. Diagram of heat treatment of specimens of A and B variants of the investigated Fe-Ni alloy

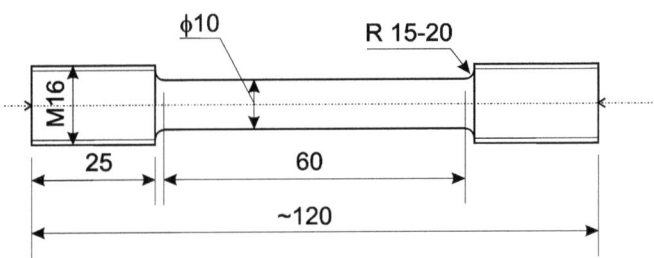

Fig. 6. Sample for static tensile tests at room and elevated temperatures

Low-cycle fatigue tests were carried out at room temperature and a temperature of 600°C using a servo-hydraulic system, MTS-810. The tests were carried with the servo-hydraulic machine being controlled by strain (the so-called fixed control) for the range of total strain $\Delta\varepsilon_t$ from 0.6 to 1.4 %. In a sinusoidal deformation cycle, an average strain rate $\dot{\varepsilon} = 1.5 \times 10^{-5}\,\text{s}^{-1}$ was applied. The number of cycles until failure of specimen N_f was assumed to be the criterion of the investigated materials' durability [62÷66]. Cylindrical specimens with a diameter $d_0 = 12$ mm and measuring length $l_0 = 30$ mm were used for tests (Fig. 7).

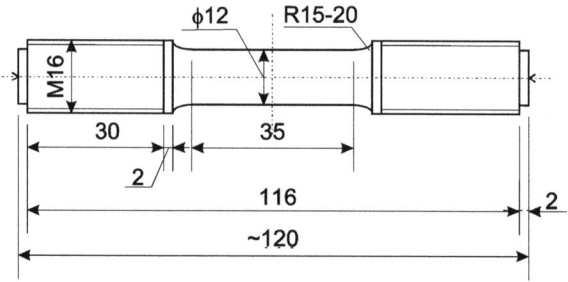

Fig. 7. Sample for low-cycle fatigue tests at room and elevated temperatures

Heating of the specimens examined at 600°C was performed with the use of Lepel induction heater (power 12.5 kVA, frequency 450 kHz) and a cylindrical inductor with a shape selected so as to obtain a possibly uniform temperature distribution along the specimen length limited by a sensor base. During the test, the specimens temperature was controlled by thermocouples (PtRh-Pt) welded to the specimen surface to an accuracy of ±5°C.

Shortened creep tests of specimens made of the investigated Fe-Ni alloy were conducted in the creep laboratory of The Institute for Ferrous Metallurgy in Gliwice. The tests were performed in compliance with standard PN-76/H-04330 in a temperature range of 650÷750°C and at stresses between 70 and 340 MPa [67]. The tests were carried out in single-specimen six-stand testing machines of maximum load of 50 kN, adjusted to measure elongation during the test. Specimens with diameter d_0 = 5 mm and measuring length l_0 = 50 mm were used for creep tests (Fig. 8).

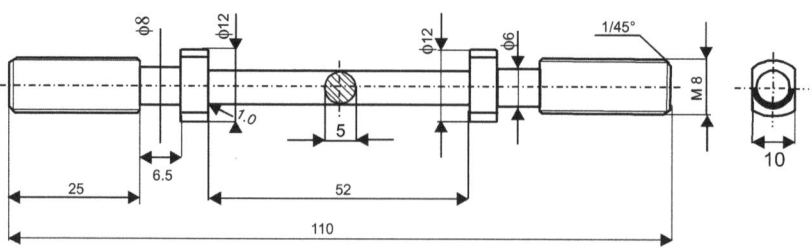

Fig. 8. Sample for shortened creep tests at elevated temperatures

The tests were carried out until specimen failure (in the time range from 60 to 1301 h) and elongation of the specimen was measured during the test. After failure, the final elongation and reduction of area of the specimen were measured at room temperature. For cognitive and comparative purposes, creep tests were conducted on alloy specimens after solution heat treatment and single-stage ageing (variant A), and after solution heat treatment followed by two-stage ageing (variant B). Specimens of variant A for the creep tests were marked with subsequent numbers A1÷A9, whereas specimens of variant B, with subsequent numbers B1÷B9. Detailed date regarding the method of marking the specimens of Variants A and B and the parameters of creep tests conducted for particular specimens are provided in Table 8.

Specimens structural tests were conducted on a Reichert MeF-2 light microscope. The surface of specimens with diameters of 10 and 12 mm was initially ground on a disc grinder and next, on waterproof abrasive papers with graining of 80÷2000. Final surface processing consisted of polishing with diamond paste on a semi-automatic Struers grinding machine. The specimens were etched using a reagent with the following composition: 54 cm^3 of hydrofluoric acid (HF), 8 cm^3 of nitric acid (HNO_3) and 38 cm^3 of distilled water.

Tests of the specimens substructure were carried out using a thin foil technique on a Jeol transmission electron microscope, JEM-2000 FX, at accelerating voltage of 160 kV. The discs for thin foils with a diameter of 3.0 mm and thickness of about 0.5 mm were cut out from a previously prepared shaft, 3.0 mm in diameter, by means of a Struers cutting-off machine, Acutom. The discs were initially ground with waterproof abrasive papers until the thickness of ca. 0.05 mm was obtained. The so obtained discs were then thinned via two-sided jet electrolytic polishing method in a Tenupol-3 device of Struers manufacture. A company brand reader A-8 was used (for alloys with a Fe matrix) cooled down to a temperature 15°C at polishing voltage of 80 V.

Fractographic tests fractures of specimens after static tensile tests, fatigue analysis and creep tests were carried out on a Jeol JSM-35 scanning microscope. Structural observations were conducted on fractures specimens produced after static tensioning tests, during the low-cycle fatigue tests and after creep tests. The purpose of fractographic tests was to evaluate the nature of fractures which formed on the specimens during fatigue or creep and to analyze plastic properties of the material after both variants of heat treatment.

5. Microstructure and mechanical properties of Fe-Ni superalloy

The results of specimens microscope observations of the Fe-Ni alloy after both variants of heat treatment are presented in Figs. 9 and 10. In both cases, the initial alloy structure demonstrated an austenitic matrix with a diversified grain size and with numerous twin systems as well as particles of primary and secondary precipitates. By comparing both of the Fe-Ni alloy structures, it can be assumed that in the alloy after 2-stage ageing (variant B), a higher fraction of secondary phase particle precipitates is observed on grain boundaries in relation to 1-stage ageing (variant A).

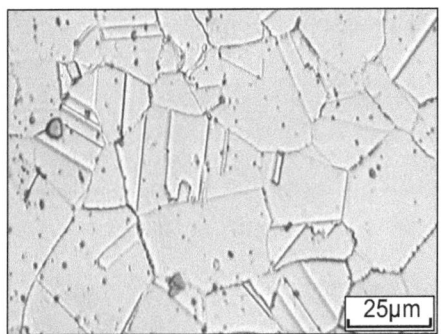 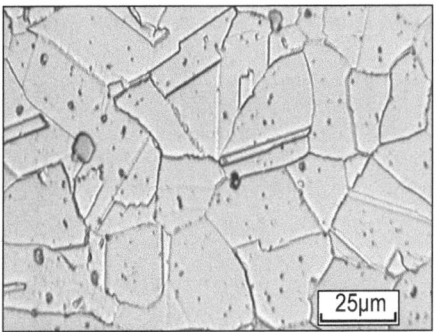

Fig. 9. Alloy microstructure after solution heat treatment and ageing according to variant A. Austenite with a diversified grain size, with primary and secondary precipitates

Fig. 10. Alloy microstructure after solution heat treatment and ageing according to variant B. Austenite with a diversified grain size, with primary and secondary precipitates

This finding is corroborated by the results of reserach on the Fe-Ni alloy substructure conducted using a transmission electron microscope (Figs. 11 and 12). It has been found that the precipitation process in the alloy substructure for variant A took place mainly within the matrix, where a characteristic "tweed-like" contrast connected with the occurrence of coherent particles of the intermetallic phase type γ'-Ni$_3$(Al,Ti) was identified

(Figs. 11, 12). As for variant B, the precipitation process of secondary phase particles took place both within the matrix and along the grain boundaries (Fig. 12). Early stages of type γ' phase precipitates were observed in the matrix, whereas within the area of grain boundaries, the occurrence of $M_{23}C_6$ carbide lamellae and lenticular particles of the G ($Ni_{16}Ti_6Si_7$) intermetallic phase [39] were observed.

The static tensile test conducted on Fe-Ni alloy specimens at room and elevated temperatures were supplement to the creep test conducted in parallel. For cognitive and comparative purposes, the tests were conducted on alloy specimens after solution heat treatment and single-stage ageing (variant A), and after solution heat treatment followed by two-stage ageing (variant B). The obtained results of tests of mechanical properties of the Fe-Ni alloy at room and elevated temperatures are presented in Table 3.

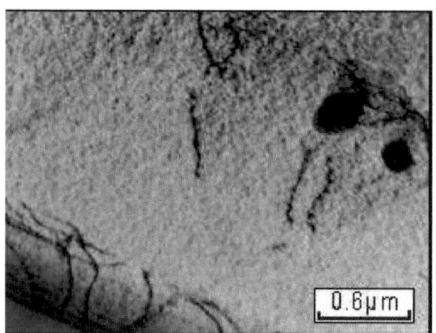 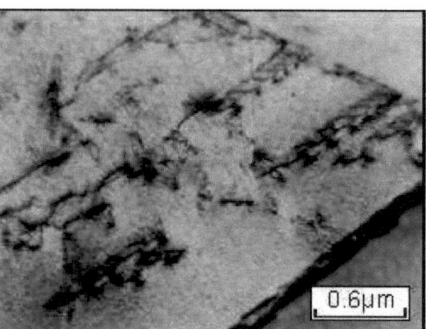

Fig. 11. Alloy substructure after heat treatment according to variant A. Coherent precipitates of phase γ' and lenticular particles of phase G in the matrix

Fig. 12. Alloy substructure after heat treatment according to variant B. Coherent precipitates of phase γ' in the matrix and $M_{23}C_6$ carbide lamellae, and phase G particles on grain boundary

Based on the provided results of tests at room temperature (20°C), it can be seen that specimens of variant B, i.e. those after 2-stage ageing, demonstrated better strength properties (Y.S = 761 MPa, T.S = 1097 MPa). Specimens of variant A (after 1-stage ageing) demonstrated a little worse strength properties (Y.S = 702 MPa, T.S = 1021 MPa) with their plastic properties being comparable.

Table 3. Mechanical properties of the Fe-Ni alloy after ageing according to variants A and B at room and elevated temperatures.

Variant of ageing	Test temperature [°C]	Y.S [MPa]	T.S [MPa]	EL. [%]	R.A. [%]
A	20	701	1021	27	48
B		761	1097	26	46
A	600	632	802	12	39
B		698	879	11	37
A	650	611	708	11	29
B		639	762	10	25
A	700	513	560	10	38
B		473	531	10	35
A	750	363	421	27	59
B		368	433	22	54

Also, tests at elevated temperature, in the range of 600÷750°C, showed that specimens of variant B (Table 2) were characterized with better strength properties (Y.S = 699÷368 MPa, T.S = 879÷433 MPa), with their plastic properties being a little worse (EL. = 10÷22%). Specimens of variant A in an analogical range of test temperatures demonstrated a little worse strength properties (Y.S = 632÷363 MPa, T.S = 802÷421 MPa) with their plastic properties being a little higher (EL. = 10÷27%).

Fractographic observations were conducted on the Fe-Ni specimen fractures after tensioning at room temperature and an increased temperature of 600°C. The results of studies for both variants of heat treatment are presented in Figs. 13÷16. In variant A specimens after 1-stage ageing and tensioning at room temperature, a transcrystalline ductile fracture with traces of significant plastic deformation was found (Fig. 13). In variant B specimens after 2-stage ageing and analogical tensioning, a similar type of ductile fracture was observed (Fig. 14). Also, tests at elevated temperature 600°C, showed a transcrystalline ductile fractures with traces of plastic deformation was found (Figs. 15,16). A specimen heat treated according to variant B, i.e. after 2-stage ageing, was characterized with a higher fraction of intergranular cracks (Fig. 16).

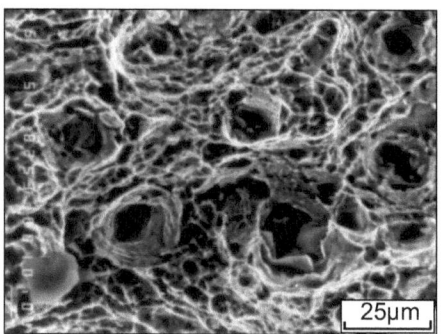

Fig. 13. Variant A specimen fracture after tensioning at a temperature of 20°C. Transcrystalline ductile fracture with traces of significant plastic deformation

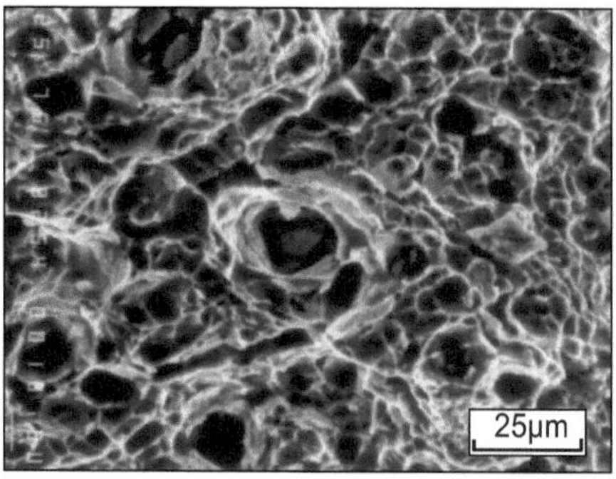

Fig. 14. Variant B specimen fracture after tensioning at a temperature of 20°C. Transcrystalline ductile fracture with traces of significant plastic deformation

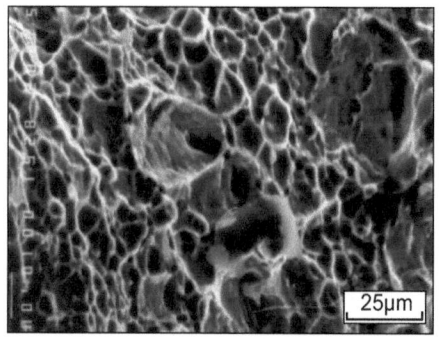
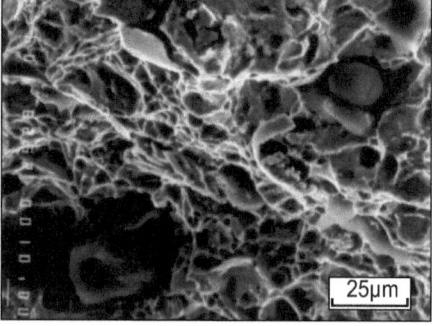

Fig. 15. Variant A specimen fracture after tensioning at a temperature of 600°C. Transcrystalline ductile fracture with traces of plastic deformation

Fig. 16. Variant B specimen fracture after tensioning at a temperature of 600°C. Transcrystalline ductile fracture with traces of plastic deformation and intergranular cracks

Based on the test results obtained, it is possible to affirm that specimens made of the examined alloy subjected to 2-stage ageing (variant

B) were characterized with stronger strengthening at both room and elevated temperatures. In turn, specimens of the examined alloy subjected to 1-stage ageing (variant A) were characterized with a little higher plasticity at room temperature and at elevated temperatures. Higher strength-related properties with the slightly lower plastic properties of the specimens after 2-stage ageing can be accounted for by stronger strengthening of grain boundaries and the zones near boundaries through precipitation of $M_{23}C_6$ carbides and phase G ($Ni_{16}Ti_6Si_7$) [39].

6. Fatigue behaviour of Fe-Ni superalloy

The results of fatigue tests conducted at temperatures of 20°C on Fe-Ni alloy specimens heat treated according to variants A and B are provided in Table 4 and presented in Figs. 17 and 18. During the low-cycle fatigue tests for individual ranges of total strain $\Delta\varepsilon_t$ (0.6÷1.4%), the values of amplitudal stress σ_a were determined depending on the number of cycles N.

Table 4. Results of low-cycle fatigue tests of the Fe-Ni alloy specimens, variants A and B at a temperature of 20°C.

Variant of ageing	Ranges of strain				σ_{an} [MPa]	N_f
	$\Delta\varepsilon_t$		$\Delta\varepsilon_e$	$\Delta\varepsilon_p$		
A	0.6%	0.006	0.0054	0.0006	593	23770
	0.8%	0.008	0.0060	0.0020	634	13520
	1.0%	0.010	0.0064	0.0036	677	9064
	1.2%	0.012	0.0068	0.0052	717	5820
	1.4%	0.014	0.0071	0.0069	744	3120
B	0.6%	0.006	0.0055	0.0005	611	20460
	0.8%	0.008	0.0064	0.0016	675	11740
	1.0%	0.010	0.0068	0.0032	712	6120
	1.2%	0.012	0.0072	0.0048	725	4320
	1.4%	0.014	0.0074	0.0066	737	2790

Based on those data, graphs of cyclic softening were built and the values of saturation stress σ_{an} were determined for the studied alloy. As results from the low-cycle tests conducted at a temperature of 20°C, the specimens subjected to ageing according to variant A show higher fatigue durability, while the specimens aged according to variant B demonstrate higher stress saturation (Figs. 17, 18). In both variants of heat treatment, the Fe-Ni alloy is characterized by cyclic softening in the low-cycle fatigue conditions.

The results of low-cycle fatigue tests carried out at an elevated temperature of 600°C are provided in Table 5 and presented in Figs.19 and 20.

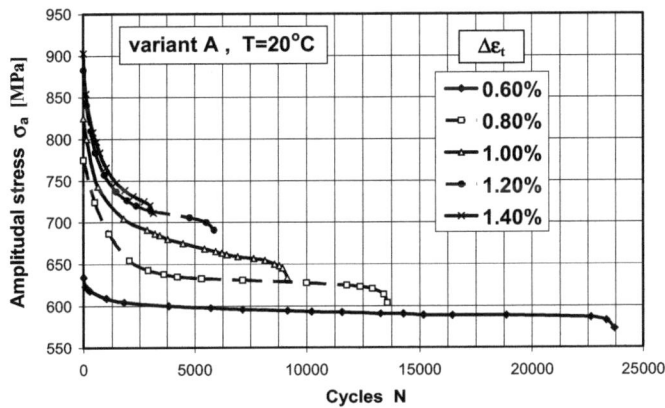

Fig. 17. Cyclic softening curves of the Fe-Ni alloy for variant A
at a temperature 20°C

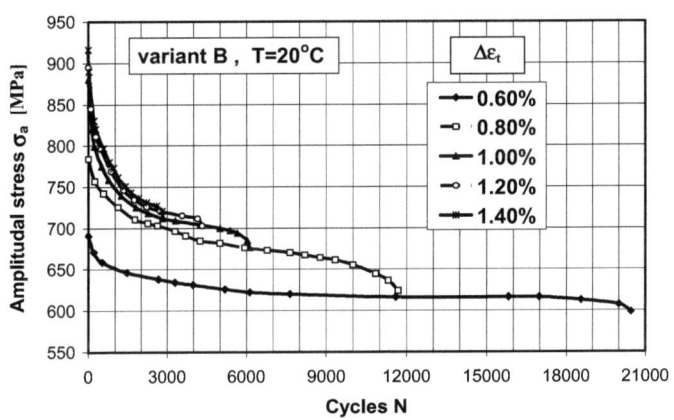

Fig. 18. Cyclic softening curves of the Fe-Ni alloy for variant B
at a temperature 20°C

Table 5. Results of low-cycle fatigue tests of the Fe-Ni alloy specimens, variants A and B at a temperature of 600°C.

Variant of ageing	Ranges of strain			σ_{an} [MPa]	N_f
	$\Delta\varepsilon_t$	$\Delta\varepsilon_e$	$\Delta\varepsilon_p$		
A	0.8%	0.008	0.0069	0.0011	585	1440
	1.2%	0.012	0.0072	0.0048	610	432
B	0.8%	0.008	0.0064	0.0016	540	500
	1.2%	0.012	0.0071	0.0049	600	310

31

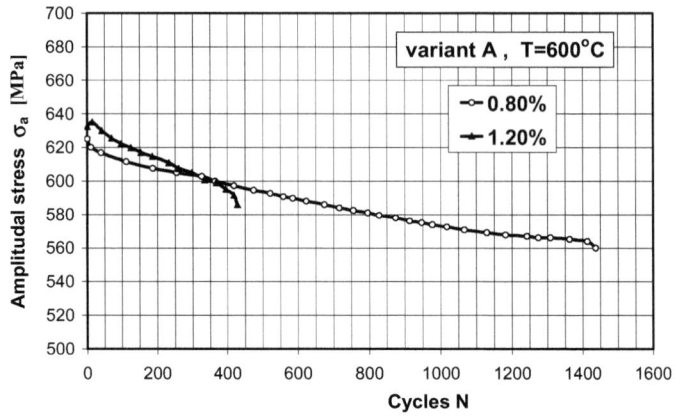

Fig. 19. Cyclic softening curves of the Fe-Ni alloy for variant A at a temperature 600°C

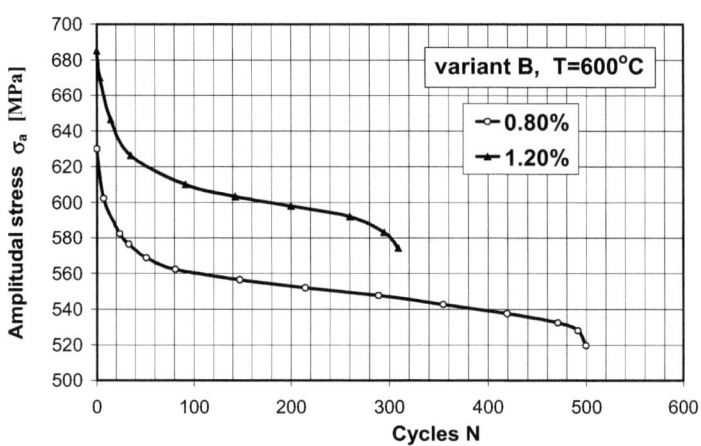

Fig. 20. Cyclic softening curves of the Fe-Ni alloy for variant B at a temperature 600°C

Based on the low-cycle tests conducted at a temperature of 600°C it was found that similarly to the tests at room temperature, the specimens aged according to variant A are characterized by higher fatigue durability (Fig. 19). The specimens subjected to 1-stage ageing demonstrate higher stress saturation σ_{an}. For both variants of heat treatment, the alloy becomes cyclically softened in low-cycle fatigue conditions. Particularly intense softening is observed in the alloy subjected to heat treatment following variant B, which decreases its operational usability in the conditions of cyclic fatigue at increased temperatures (Fig. 20).

The ranges of plastic strain $\Delta\varepsilon_p$ and elastic strain $\Delta\varepsilon_e$, and their corresponding stress range $\Delta\sigma$, were determined on the basis of a hysteresis loop recorded in the course of the testing. The results obtained were used to elaborate a fatigue durability graph of the studied alloy. The fatigue durability values for the Fe-Ni alloy at room temperature were described by the Smith, Hirschberg and Manson dependence [68]:

$$\Delta\varepsilon_t = \Delta\varepsilon_p + \Delta\varepsilon_e = M \cdot N_f^z + \frac{G}{E} \cdot N_f^v \qquad (1)$$

where: M, G, E, z, v – material constants.

The results of the Fe-Ni alloy fatigue durability at room temperature are provided in Table 6 and illustrated in Figs. 21 and 22.

Table 6. Mathematical models of the Fe-Ni alloy specimens fatigue durability for variant A and B at room temperature.

Variant of ageing	$\Delta\varepsilon_p = M \cdot N_f^z$		$\Delta\varepsilon_e = (G/E) \cdot N_f^v$	
	M	z	G/E	v
A	126.0	-1.18	0.0217	-0.136
B	188.4	-1.27	0.0240	-0.144

An analysis of the Fe-Ni alloy fatigue durability graphs at room temperature has shown that for both ageing variants, A and B, the intersection point N_t of graphs $\Delta\varepsilon_e = f(N_f)$ and $\Delta\varepsilon_p = f(N_f)$ is located in the low-cycle range, i.e. 4000 and 3000 cycles, respectively (Figs. 21, 22). This testifies to the fact that the cyclic deformation process of the alloy was proceeding with a dominant participation of the elastic component $\Delta\varepsilon_e$ within the complete strain ranges $\Delta\varepsilon_t$ assumed for the tests (Table 4). In both of the studied ageing variants of the Fe-Ni alloy, the resistance to plastic deformation depends mainly on its strength-related properties.

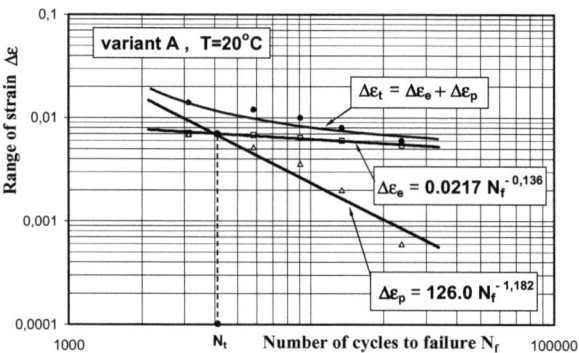

Fig. 21. Fatigue durability graphs of the Fe-Ni alloy for variant A at a temperature 20°C

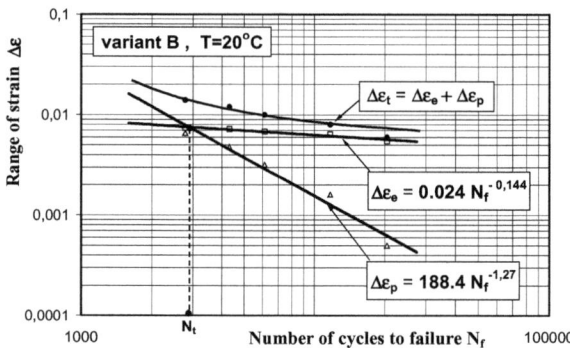

Fig. 22. Fatigue durability graphs of the Fe-Ni alloy for variant B at a temperature 20°C

A comparison of the influence of the Fe-Ni alloy both ageing variants on fatigue durability (N_f) at room and increased temperatures is presented in Figs. 23 and 24. As can be seen from the data provided, both at a room temperature and at a temperature increased to 600°C, the alloy fatigue durability was higher for variant A compared to variant B.

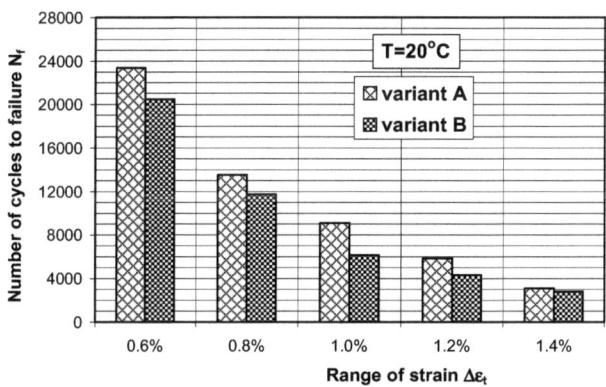

Fig. 23. Comparison of fatigue durability of ageing variants A and B for the Fe-Ni alloy at room temperature

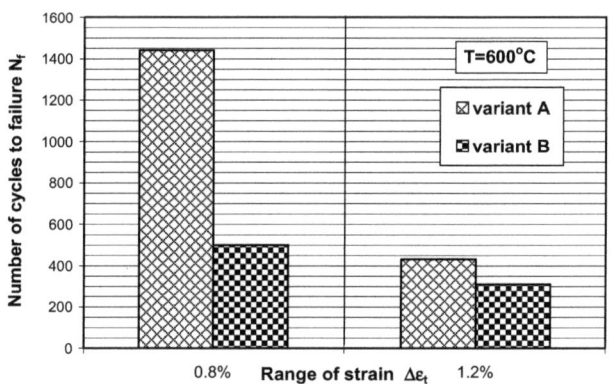

Fig. 24. Comparison of fatigue durability of ageing variants A and B for the Fe-Ni alloy at an increased temperature of 600°C

Having obtained the saturation stress values σ_{an} for amplitudal plastic strain ε_p, mathematical models (2) of cyclic alloy deformation were devised, as given in Table 7 and presented graphically in Fig. 25. Also, a cyclic strength coefficient (K') and a cyclic weakening exponent (n') were determined for the studied alloy [68]:

$$\sigma_{an} = K' \cdot \left[\frac{\varepsilon_p}{2} \right]^{n'} \tag{2}$$

where: K' – cyclic strength coefficient, n' – cyclic softening exponent.

Table 7. Values of coefficient (K') and exponents (n') for the Fe-Ni alloy deformation curves at 20°C of variants A and B.

Variant of ageing	ε_p	σ_{an} [MPa]	K' [MPa]	n'
A	0.0003	593	1233.2	0.092
	0.0010	634		
	0.0018	677		
	0.0026	717		
	0.0034	744		
B	0.00025	611	1131.7	0.074
	0.0008	675		
	0.0016	712		
	0.0024	725		
	0.0033	737		

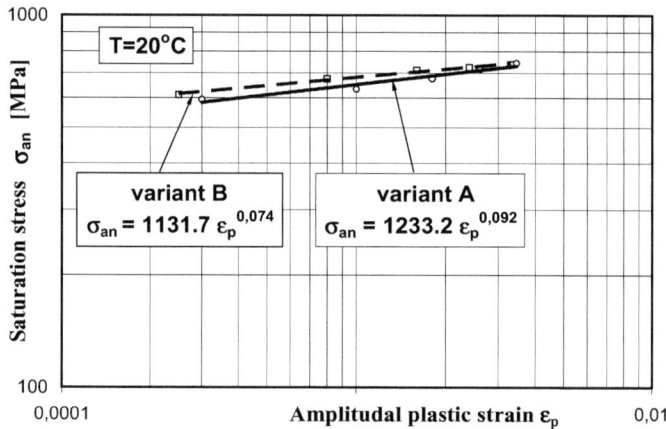

Fig. 25. Cyclic deformation graph of the specimens for ageing variants A and B of the Fe-Ni alloy at room temperature

The results of microscopic observations conducted on Fe-Ni specimens after fatigue tests at room and elevated temperatures are presented in Figs. 26÷33. After low-cycle fatigue tests at a temperature of 20°C within the range of complete deformation ($\Delta\varepsilon_t$ = 0.8%), occurrence of deformed austenite grains with slip lines and bands, and secondary phase particles, were detected in the alloy microstructure (Figs. 26, 27). An increase in total deformation to the value of $\Delta\varepsilon_t$ = 1.2% was accompanied in the alloy microstructure by increasing density of the slip lines and bands inside the deformed austenite grains (Figs. 28, 29). No significant differences were found in the microstructure of the specimens after low-cycle fatigue, heat treated according to variants A and B.

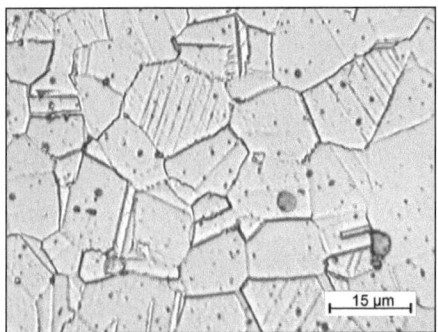

Fig. 26. Variant A specimen microstructure after fatigue tests ($\Delta\varepsilon_t$ = 0.8%) at room temperature. Slip lines and bands in the interior of deformed grains

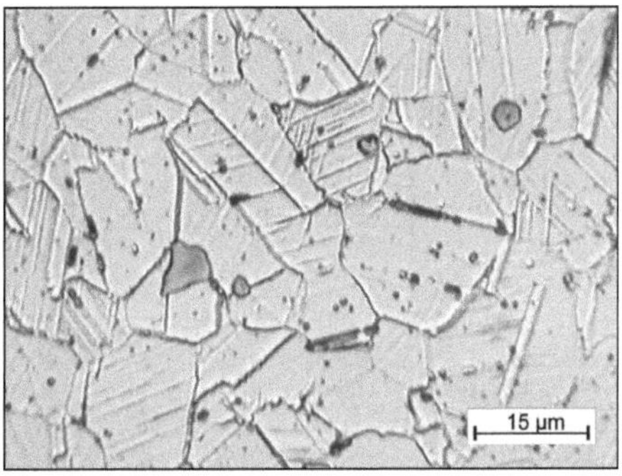

Fig. 27. Variant B specimen microstructure after fatigue tests ($\Delta\varepsilon_t$ = 0.8%) at room temperature. Slip lines and bands in the interior of deformed grains

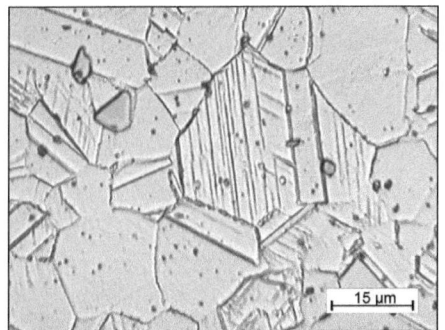
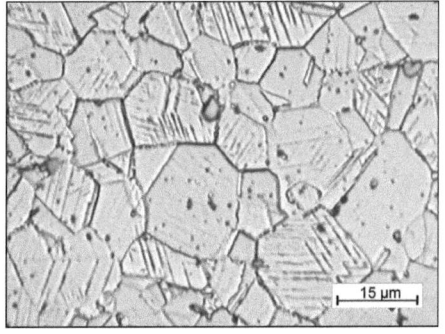

Fig. 28. Variant A specimen microstructure after fatigue tests ($\Delta\varepsilon_t$ = 1.2%) at room temperature. Slip lines and bands in the interior of deformed grains

Fig. 29. Variant B specimen microstructure after fatigue tests ($\Delta\varepsilon_t$ = 1.2%) at room temperature. Slip lines and bands in the interior of deformed grains

In the fatigue-deformed specimens within the range of deformation $\Delta\varepsilon_t$ = 0.8÷1.2% at temperature 600°C, deformed austenite grains with secondary phase particles only were found in the microstructure (Figs. 30÷33). In the case of both initial heat treatment variants no slip lines or bands were observed in the microstructure, which testifies to the progressing dynamic recovery during low-cycle fatigue at an elevated temperature. Also, no significant differences were found in the microstructure of the specimens after low-cycle fatigue at 600°C, heat treated according to variants A and B.

Fig. 30. Variant A specimen microstructure after fatigue tests ($\Delta\varepsilon_t$ = 0.8%) at a temperature of 600°C. Traces of deformation at austenite grain boundaries

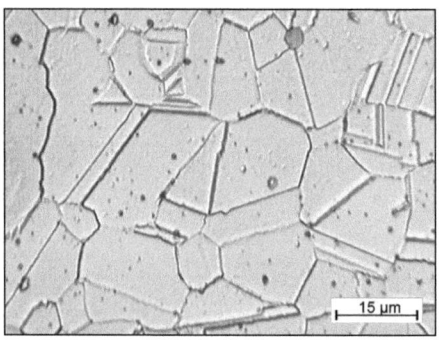

Fig. 31. Variant B specimen microstructure after fatigue tests ($\Delta\varepsilon_t$ = 0.8%) at a temperature of 600°C. Traces of deformation at austenite grain boundaries

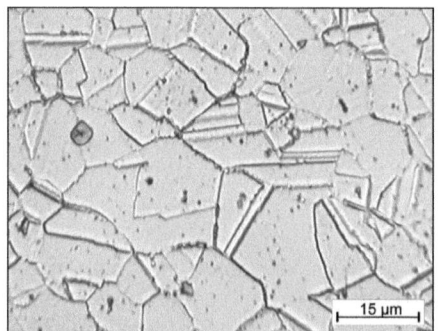

Fig. 32. Variant A specimen microstructure after fatigue tests ($\Delta\varepsilon_t$ = 1.2%) at a temperature of 600°C. Traces of deformation at austenite grain boundaries

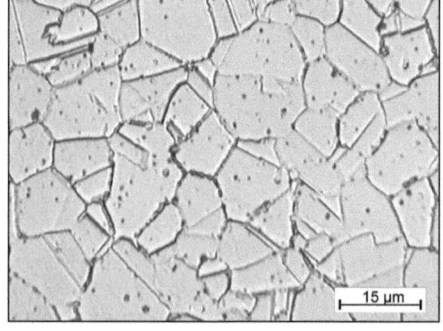

Fig. 33. Variant B specimen microstructure after fatigue tests ($\Delta\varepsilon_t$ = 1.2%) at a temperature of 600°C. Traces of deformation at austenite grain boundaries

Fractographic observations were conducted on the Fe-Ni specimen fractures after low-cycle fatigue at room temperature and an increased temperature of 600°C. The results of studies for both variants of heat treatment are presented in Figs. 34÷41. After low-cycle fatigue tests conducted until total strain $\Delta\varepsilon_t$ = 0.8% at room temperature, the specimens demonstrated a certain diversification in terms of the obtained fractures morphology (Figs. 34, 35). In variant A specimens, a typical fatigue fracture with characteristic fatigue stripes and fraction of secondary intergranular cracks were observed (Fig. 34). In the case of variant B specimens, the fatigue fracture was of a cleavage type, with a fraction of secondary intergranular cracks on grain boundaries (Fig. 35). An increase in total deformation to the value of $\Delta\varepsilon_t$ = 1.2% was accompanied by increasing of volume fraction of transgranular cracks and traces of plastic deformation in the fatigue fracture (Figs. 36, 37). No significant differences were found in the fatigue fracture of the specimens after low-cycle fatigue, heat treated according to variants A and B.

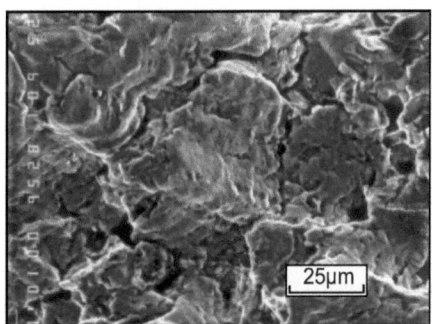 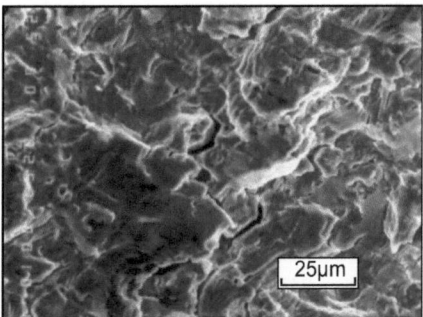

Fig. 34. Variant A specimen fatigue fracture after fatigue tests ($\Delta\varepsilon_t$ = 0.8%) at a temperature of 20°C. Fatigue fracture with fatigue stripes and secondary cracks

Fig. 35. Variant B specimen fatigue fracture after fatigue tests ($\Delta\varepsilon_t$ = 0.8%) at a temperature of 20°C. Fatigue fracture with secondary cracks on grain boundaries

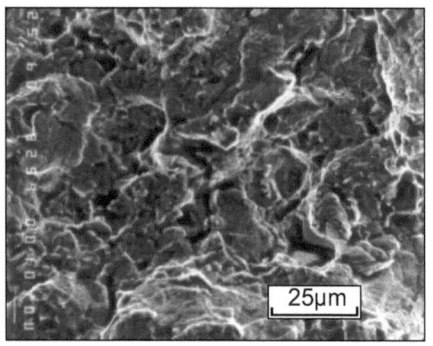 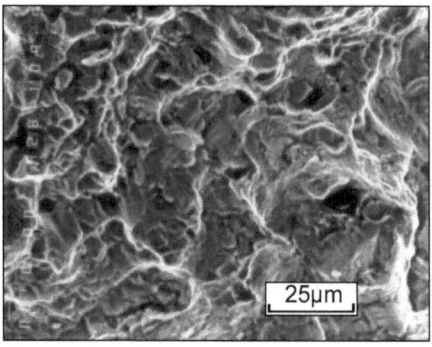

Fig. 36. Variant A specimen fatigue fracture after fatigue tests ($\Delta\varepsilon_t$ = 1.2%) at a temperature of 20°C. Mixed fatigue fracture with secondary cracks on grain boundaries

Fig. 37. Variant B specimen fatigue fracture after fatigue tests ($\Delta\varepsilon_t$ = 1.2%) at a temperature of 20°C. Mixed fatigue fracture with secondary cracks on grain boundaries

Also, after low-cycle fatigue tests conducted until total strain $\Delta\varepsilon_t$ = 0.8÷1.2% at an increased temperature of 600°C, the specimens demonstrated significant diversification in terms of the obtained fractures morphology (Figs. 38÷41). In variant A specimens, an mixed fatigue fracture with a fraction of secondary intergranular cracks were observed (Fig. 38). In the case of variant B specimens, the fatigue fracture was of cleavage type, with a fraction of secondary intergranular cracks were observed (Fig. 39). An increase in total deformation to the value of $\Delta\varepsilon_t$ = 1.2% does not cause significant changes in the fatigue fracture of samples in variant A (Fig. 40). In the case of variant B specimens, a typical intergranular fatigue fracture with secondary cracks on grain boundaries were observed (Fig. 41). The occurrence of this type of fracture in low-cycle fatigue tests at a temperature of 600°C proves the materials substantial brittleness in variant B specimens at increased temperature.

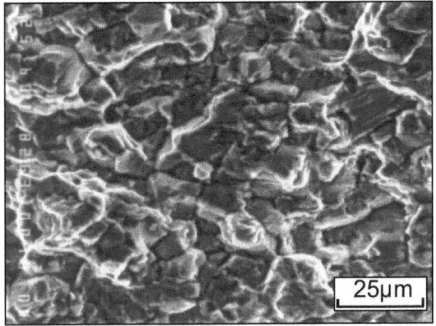

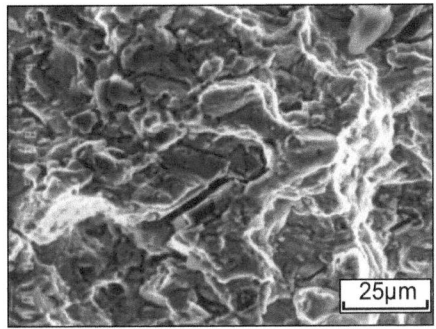

Fig. 38. Variant A specimen fatigue fracture after fatigue tests ($\Delta \varepsilon_t$ = 0.8%) at a temperature of 600°C. Mixed fatigue fracture with secondary cracks

Fig. 39. Variant B specimen fatigue fracture after fatigue tests ($\Delta \varepsilon_t$ = 0.8%) at a temperature of 600°C. Mixed fatigue fracture with secondary cracks

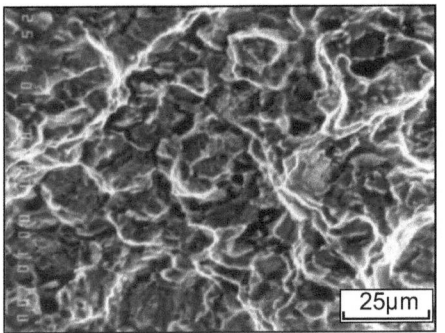

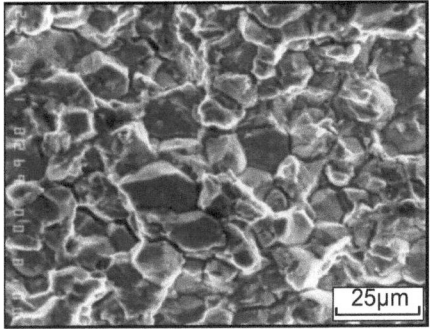

Fig. 40. Variant A specimen fatigue fracture after fatigue tests ($\Delta \varepsilon_t$ = 1.2%) at a temperature of 600°C. Mixed fatigue fracture with secondary cracks

Fig. 41. Variant B specimen fatigue fracture after fatigue tests ($\Delta \varepsilon_t$ = 1.2%) at a temperature of 600°C. Intergranular fatigue fracture with secondary cracks

7. Creep resistance of Fe-Ni superalloy

Results of shortened creep tests of specimens of variants A and B at temperatures from the range of 650÷750°C and stress of 70÷340 MPa are given in Table 8. They are also presented in a form of cumulative creep curves in consecutive Figs. 42÷44. It appears from the diagrams obtained that all creep curves have a shape characteristic of the accelerated creep stage. For a given test temperature, the shape of creep curves significantly depends on the value of the stress applied. A comparison of the alloy creep curves at a temperature of 650°C gave quite diversified results, depending on the value of the stress applied (Fig. 42). At high stresses, i.e. 340 and 260 MPa, specimens of variant B demonstrated a slower deformation rate, whereas at a lower stress (240 MPa), specimens of variant A showed a slower deformation rate.

At a higher test temperature, 700°C, a similar diversity of results concerning the creep resistance was obtained (Fig. 43). At higher stresses, of 240 and 180 MPa, specimens of variant B demonstrated a lower deformation rate. In turn, at a lower stress (150 MPa), specimens of variant A showed better creep resistance.

A comparison of creep curves for the alloy at the highest tested temperature, i.e. 750°C, also yields diversified results, depending on the value of the stress applied (Fig. 44). At high stresses (σ = 120 MPa), a specimen of variant B demonstrated a slower deformation rate. Whereas at a lower stress (σ = 80 MPa), a specimen of variant A showed a slower deformation rate. At the lowest stress (σ = 70 MPa), the deformation rates for specimens of variants A and B were similar.

Table 8. Specification of specimens and results of shortened creep tests of the Fe-Ni alloy after ageing for variants A and B.

Sign of sample	Temperature [°C]	Stress [MPa]	Time to rupture [h]	Elongation [%]	Reduction of area [%]
A1	650	340	70	6	30
B1	650	340	132	5	28
A2	650	260	649	9	28
B2	650	260	709	7	11
A3	650	240	1014	7	15
B3	650	240	1025	7	16
A4	700	240	60	15	20
B4	700	240	99	15	19
A5	700	180	941	22	43
B5	700	180	1009	15	36
A6	700	150	1186	21	39
B6	700	150	1165	22	42
A7	750	120	120	28	73
B7	750	120	170	28	75
A8	750	80	789	33	77
B8	750	80	692	35	42
A9	750	70	1284	27	48
B9	750	70	1301	32	54

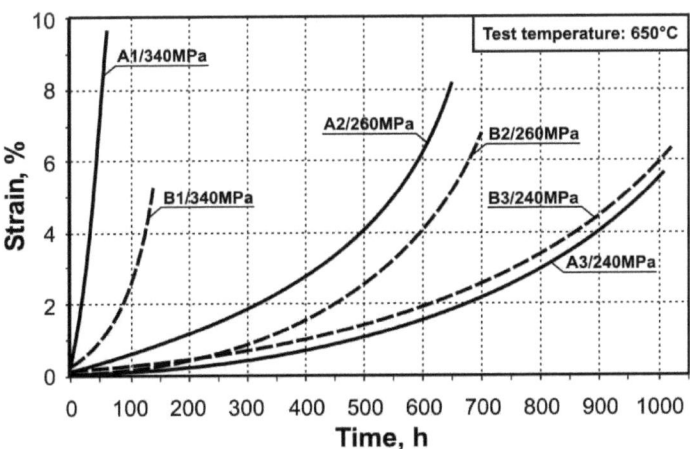

Fig. 42. Comparison of creep curves of specimens of variants A and B at a temperature 650°C and stresses: 340, 260 and 240 MPa

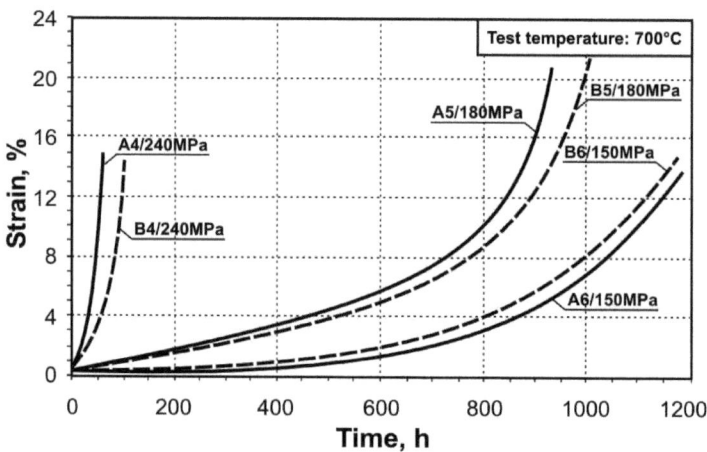

Fig. 43. Comparison of creep curves of specimens of variants A and B at a temperature 700°C and stresses: 240, 180 and 150 MPa

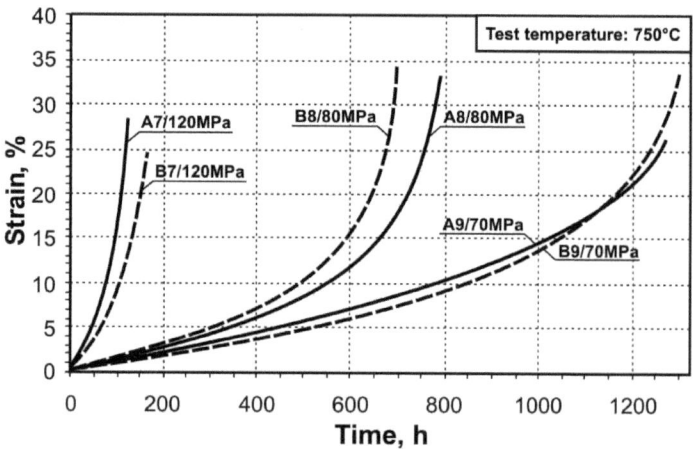

Fig. 44. Comparison of creep curves of specimens of variants A and B at a temperature 750°C and stresses: 120, 80 and 70 MPa

Based on the obtained shortened creep curves of specimens of the tested Fe-Ni alloy, the creep test results were then extrapolated. The extrapolation was performed using a graphical method [69,70] for creep times equal 100, 1000 and 10000 h, at the examined test temperatures of 650, 700 and 750°C. Results of extrapolation of the creep tests for variants A and B are given in Table 9 and presented graphically in Fig. 45. As can be seen from the results obtained, at the lowest creep temperature of 650°C and in the range of short and medium test time of 100÷1000 h, higher creep resistance was exhibited by specimens of variant B. In turn, at the longest analyzed creep time – 10000 h – specimens of variant A demonstrated better temporal strength. At higher test temperature of 700°C and short creep time of 100 h, specimens of variant B demonstrated higher creep resistance. In turn, at longer test time of 1000÷10000 h, specimens heat treated according to variant A demonstrated better temporal strength.

Table 9. Extrapolated values of temporal creep strength of specimens of variant A and B of Fe-Ni alloy for test time 100, 1000 and 10000 h.

Time of test creep [h]	Creep strength [MPa] at temperature:					
	650°C		700°C		750°C	
	Variant A	Variant B	Variant A	Variant B	Variant A	Variant B
100	327	353	226	239	124	132
1000	242	245	167	150	75	74
10000	157	138	109	61	27	16

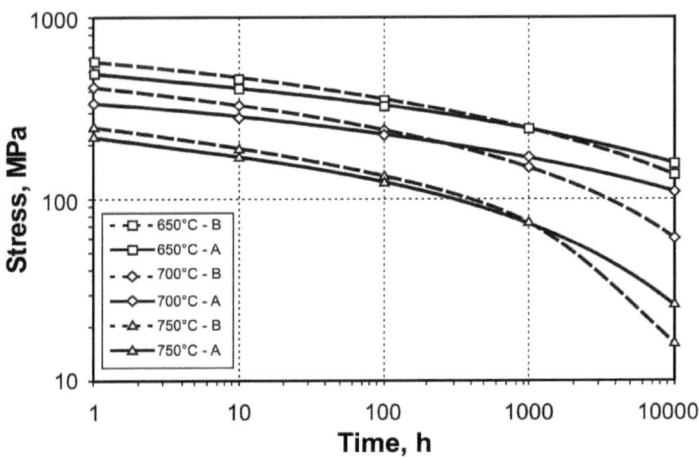

Fig. 45. Extrapolation of creep test results at the temperature of 650, 700 and 750°C for specimens of Fe-Ni alloy after ageing according to variants A and B

45

Also at the highest test temperature of 750°C and short creep time of 100 h, specimens of variant B demonstrated higher creep resistance. At the average test time of 1000 h, specimens of variants A and B obtained similar temporal strengths. At the longest analyzed creep time – 10000 h – specimens of variant A reached significantly higher creep resistance in comparison with specimens of variant B.

The results of microscopic observations conducted on Fe-Ni specimens in variants A and B after selected creep tests are presented in Figs. 46÷51. After creep tests carried out at a temperature of 650°C and stress of 240 MPa, grain nucleation and creep pores coalescence were detected at austenite grain boundaries in the alloy microstructure (Figs. 46, 47). In the austenite matrix and at boundary areas the occurrence of spheroidal particles of phase γ' - $Ni_3(Al,Ti)$, lenticular particles of phase G ($Ni_{16}Ti_6Si_7$) and lamellae of phase η (Ni_3Ti) was observed [39]. No significant differences were found in the microstructure of the specimens after creep, heat treated according to variants A and B.

An increase in the temperature of creep tests up to 700°C at stress of 150 MPa intensified decohesion and overageing in the alloy microstructure during the exposure (Figs. 48, 49). Creep microcracks appeared at austenite grain boundaries, which developed with a particular intensity at contact points of the neighbouring grain boundaries. Distinct traces of alloy overageing were observed in the austenitic matrix in the form of coagulation of phase γ' particles and formation of phase η lamellae. No significant differences were found in the microstructure of the specimens after creep in variants A and B.

At the highest applied creep temperature of 750°C and stress of 70 MPa, advanced decohesion and overageing processes were found to take place in the alloy microstructure (Figs. 50, 51). Large intergranular creep cracks developed along grain boundaries and in the grain boundary contact points. An advanced overageing stage was detected in the austenite matrix, connected with the $\gamma' \rightarrow \eta$ transition and formation of a large number of intercrystalline lamellae of phase η in a Widmanstätten pattern [1,27,39]. No significant differences were found in the microstructure of the specimens after creep, heat treated according to variants A and B.

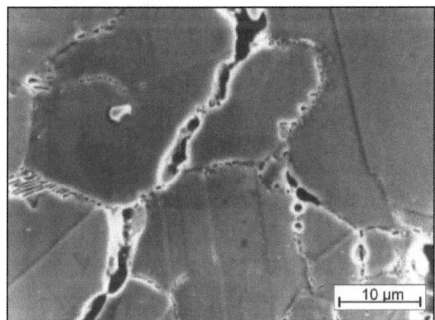 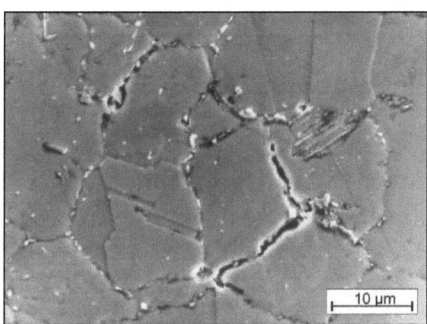

Fig. 46. Microstructure of A3 specimen after creep test at 650°C/240 MPa. Nucleation and coalescence of creep pores at austenite grain boundaries

Fig. 47. Microstructure of B3 specimen after creep test at 650°C/240 MPa. Nucleation and coalescence of creep pores at austenite grain boundaries

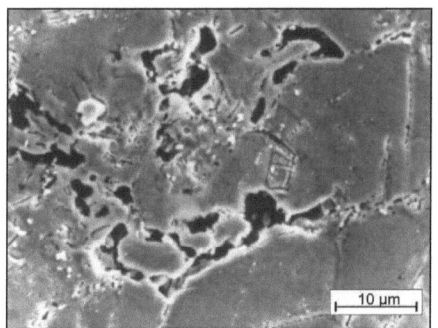 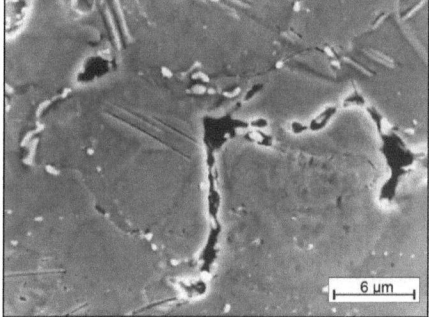

Fig. 48. Microstructure of A6 specimen after creep test at 700°C/150 MPa. Development of creep microcracks at austenite grain boundaries

Fig. 49. Microstructure of B6 specimen after creep test at 700°C/150 MPa. Development of creep microcracks at austenite grain boundaries

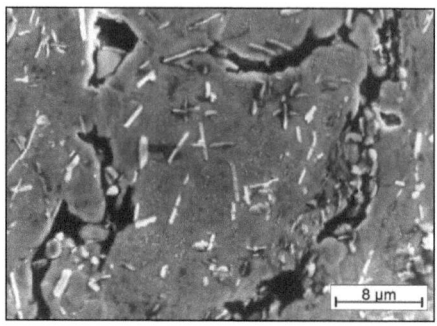

Fig. 50. Microstructure of A9 specimen after creep test at 750°C/70 MPa. Advanced stage of intergranular cracking at grain boundaries and traces of overageing in the matrix

Fig. 51. Microstructure of B9 specimen after creep test at 750°C/70 MPa. Advanced stage of intergranular cracking at grain boundaries and traces of overageing in the matrix

Structural examination with use of transmission electron microscope (TEM) was performed with thin foil technique. Specimens for thin foils were drawn transversely from the part adjacent to the fractures created during creep. The purpose of the tests was to reveal changes in the alloy substructure, formed during creep. The results of tests for the selected specimens after creep are presented in Figs. 52÷57. In the substructure of specimens subject to creep at the temperature of 650°C and stress σ = 240 MPa, dislocation clusters were revealed in boundary areas. Also, inside the matrix grains, the processes of recrystallization austenite and formation of intercrystalline lamellae of phase η were observed (Fig. 52). In the interior of matrix grains, effects of blocking the dislocation motion by dispersive γ' phase particles were observed (Fig. 53). Presence of such effects in the alloy substructure shows that creep at the temperature of 650°C is mainly of dislocation nature.

In the substructure of specimens subject to creep at the temperature of 700°C and stress σ = 150 MPa, a significantly larger density of dislocation clusters in matrix was revealed. In the deformed austenite the formation process of cellular dislocation substructure was observed (Fig. 54).

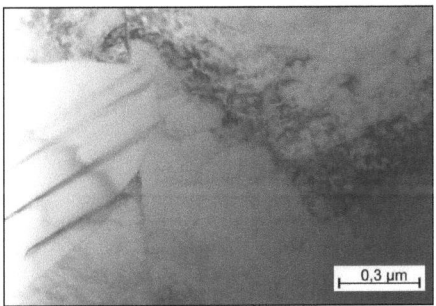 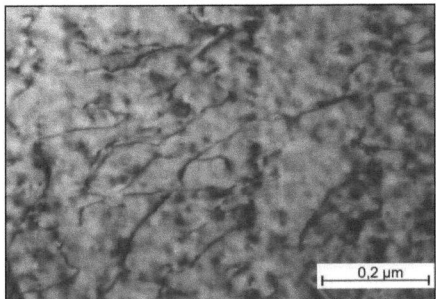

Fig. 52. Substructure of A3 specimen after creep test at 650°C/240 MPa. Recrystallized grain of austenite with lamellae of phase η and traces of deformation in the matrix

Fig. 53. Substructure of A3 specimen after creep test at 650°C/240 MPa. Blocking the dislocation motion by dispersive γ' phase particles in the matrix

In the boundary areas of the matrix, advanced stages of recovery and of coagulation of γ' phase particles were observed as well as formation of lamellae of phase η in recrystallized grains of austenite (Fig. 55). Revealing this type of substructure in the specimens analyzed shows that creep at the temperature of 700°C is of complex nature which may be defined as dislocation-diffusive.

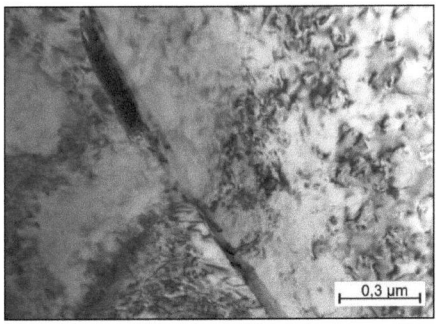 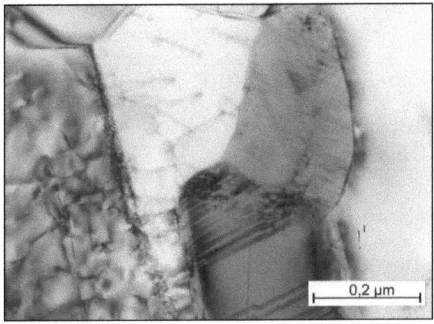

Fig. 54. Substructure of B6 specimen after creep test at 700°C/150 MPa. Formation process of cellular dislocation substructure in the matrix

Fig. 55. Substructure of B6 specimen after creep test at 700°C/150 MPa. Processes of matrix recovery and coagulation of γ' phase particles

In the substructure of specimens subjected to creep at the temperature of 750°C and stress $\sigma = 70$ MPa, areas of diverse dislocation density were detected, as well as clearly noticeable effects of overageing of the alloy connected with the $\gamma' \rightarrow \eta$ transition (Fig. 56). The formation of a number of transcrystalline and cellular η phase lamellae was accompanied with dissolution of the neighbouring γ' phase particles and decrease of dislocation density in the matrix (Fig. 57). Presence of such effects in the alloy substructure shows that creep at the temperature of 750°C is mainly of diffusive nature.

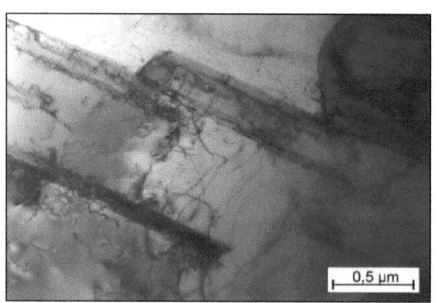 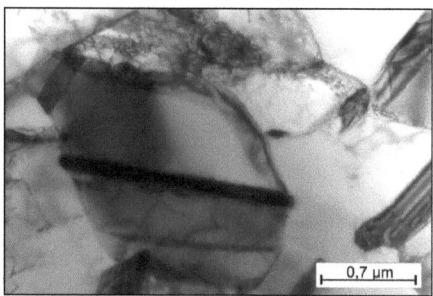

Fig. 56. Substructure of A9 specimen after creep test 750°C/70 MPa. Transcrystalline lamellae of phase η in the matrix of low dislocation density

Fig. 57. Substructure of A9 specimen after creep test 750°C/70 MPa. Recrystallized grains of austenite with lamellae of phase η within them

Fractographic tests were conducted on fractures drawn from selected specimens after creep. The purpose of the tests was to evaluate the nature of fractures formed in the specimens during creep. The results of the fractographic tests of specimens of variants A and B after creep are presented in Figs. 58÷63. In the specimens subject to creep at the temperature of 650°C and stress of 240 MPa, a predominant fraction of ductile fracture with a minor fraction of brittle intergranular cracks was revealed (Fig. 58). Specimen of variant B, i.e. after two-stage ageing, was characterized with a more cleavable nature, with a larger fraction of intergranular cracks (Fig. 59).

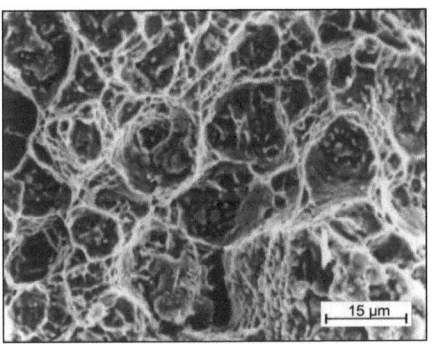 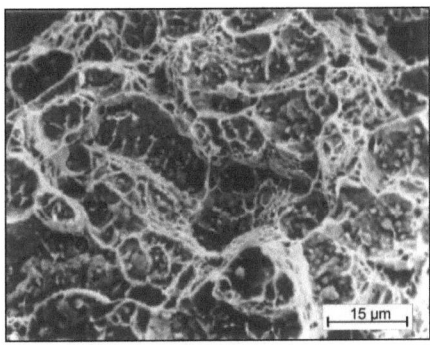

Fig. 58. Fracture of A3 specimen after creep test at 650°C/240 MPa. Mixed intergranular fracture with cleavage cracks

Fig. 59. Fracture of B3 specimen after creep test at 650°C/240 MPa. Mixed intergranular fracture with cleavage cracks

A similar type of ductile fracture with a little larger fraction of intergranular cleavage cracks was observed in specimens after creep at the temperature of 700°C and stress of 150 MPa (Fig. 60). Also in this case, a specimen heat treated according to variant B, i.e. after 2-stage ageing, was characterized with a higher fraction of brittle cracks (Fig. 61).

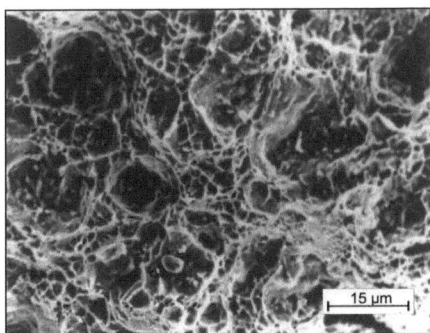 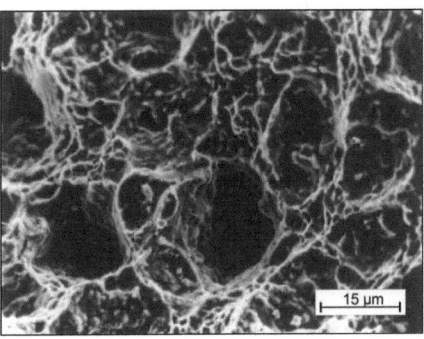

Fig. 60. Fracture of A6 specimen after creep test at 700°C/150 MPa. Mixed intergranular fracture with cleavage cracks

Fig. 61. Fracture of B6 specimen after creep test at 700°C/150 MPa. Mixed intergranular fracture with large of cleavage cracks

Also in the specimens after creep at the highest temperature of 750°C and stress of 70 MPa, a predominant fraction of ductile fracture with some fraction of cleavage cracks was revealed (Fig. 62). Specimen of variant B, was characterized with a more cleavable nature, with a larger fraction of intergranular cracks (Fig. 63).

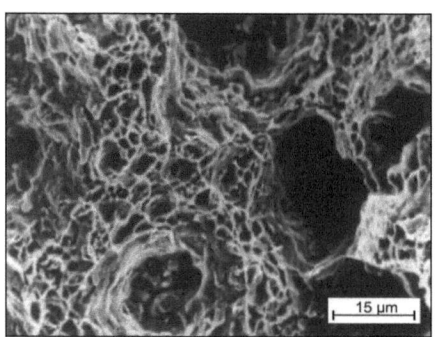

Fig. 62. Fracture of A9 specimen after creep test at 750°C/70 MPa. Mixed intergranular fracture with large of cleavage cracks

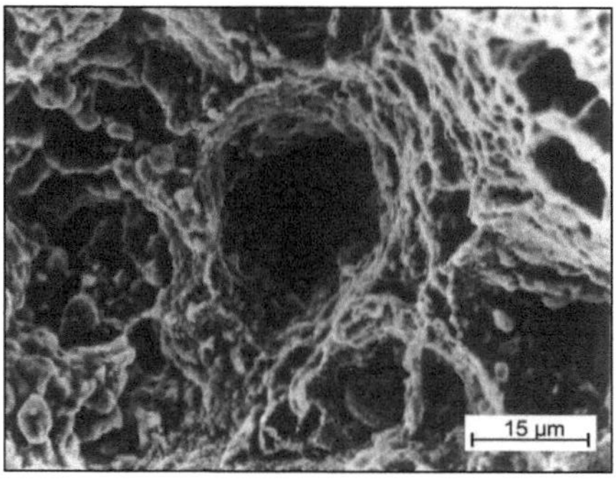

Fig. 63. Fracture of B9 specimen after creep test 750°C/70 MPa. Mixed intergranular fracture with large of cleavage cracks

In this case, on the fracture surface of both specimens tested, traces of significant oxidation were found, whereas fraction of brittle cracks for specimens of variants A and B was comparable.

8. Summary

The work analyses the influence of initial heat treatment on the mechanical properties and microstructure of the austenitic Fe-Ni alloy precipitation-strengthened with intermetallic phase of the γ' - $Ni_3(Al,Ti)$ type. Specimens of the studied alloy after solution heat treatment (980°C/2 h/water) were subjected to two ageing variants, i.e. 1-stage ageing (715°C/16 h/air) – variant A and 2-stage ageing (715°C/8 h/furnace + 650°C/8 h/air) – variant B. On heat-treated specimens according to variants A and B were performed: static tensile test at room and elevated temperatures (650÷750°C), low-cycle fatigue tests in the range $\Delta\sigma_t$ = 0.6÷1.4% at temperatures of 20 and 600°C and shortened creep tests in a range of temperature of 650÷750°C and at stresses from 70 to 340 MPa.

In the microstructure of the studied Fe-Ni alloy, for both ageing variants A and B, an austenitic matrix was detected with a diversified grain size and undissolved particles of titanium compounds, and coherent dispersion precipitates of the γ' intermetallic phase. It has been found that in the 2-stage aged alloy the precipitation process was proceeding to a greater extent along grain boundaries, where lamellar precipitates of $M_{23}C_6$ carbide and lenticular particles of phase G ($Ni_{16}Ti_6Si_7$) were identified. Such course of precipitation in the alloy heat treated according to variant B can result in both, enhanced strengthening of the areas near boundaries and in material increased brittleness in those regions.

Static tensile tests of Fe-Ni alloy, conducted at the temperature of 20°C demonstrated higher strength properties of the specimens of variant B (Y.S = 761 MPa, T.S = 1097 MPa) compared to variant A (Y.S = 701 MPa, T.S = 1021 MPa), with their plastic properties being comparable. Also, at the increased temperature of 600÷750°C, variant B specimens were characterized by higher strength properties (Y.S = 699÷368 MPa,

T.S = 879÷433 MPa) in comparison with variant A (Y.S = 632÷363 MPa, T.S = 802÷421 MPa).

The low-cycle fatigue tests proved a significant influence of the applied ageing variants A and B on the Fe-Ni alloy fatigue durability at room temperature and at 600°C. At both temperatures tested, the alloy's fatigue durability after heat treatment according to variant A was higher than the durability of the alloy treated according to variant B, whereas greater differences between the durability values (by ca. 70%) were observed at a temperature of 600°C. In the studies conducted at room temperature, the greatest differences in fatigue durability (in the range of 13÷32%) occurred in the range of total strain of 0.8÷1.2%.

The reason for lower fatigue durability at temperatures of 20 and 600°C of the heat treated specimens in variant B should be sought in a larger number of secondary phase particles precipitated on grain boundaries, which determines earlier initiation of the fatigue cracking process. This has been corroborated by the observation of the fatigue fractures morphology, where a development of intergranular cracks was found, indicating low cohesion of the grain boundaries, especially at a temperature of 600°C.

The analysis of the Fe-Ni alloy fatigue durability graphs at room temperature has shown that for the ageing variants A and B, the intersection point N_t of the graphs $\Delta\varepsilon_e = f(N_f)$ and $\Delta\varepsilon_p = f(N_f)$ is located in the low-cycle range (3000÷4000 cycles). This testifies to the fact that the cyclic deformation process of the alloy was proceeding with a dominant participation of the elastic component $\Delta\varepsilon_e$ within the total strain ranges assumed in the studies. In such conditions, the investigated alloy fatigue durability for both ageing variants was determined by its strength properties.

On the basis of the results obtained, a conclusion can be drawn that the studied Fe-Ni alloy is characterized by better material characteristics after solution heat treatment and 1-stage ageing at 715°C/16 h/air. With its slightly decreased strength properties, the alloy heat treated according to variant A shows definitely higher durability in the conditions of low-cycle fatigue, especially at elevated temperatures.

Shortened creep tests demonstrated the diversified influence of applied ageing variants A and B on the temporal strength of Fe-Ni alloy tested. It was found that in the scope of the short and medium creep time tested of ca. 100 and 1000 h at the temperature of 650÷750°C, higher creep resistance was exhibited by specimens of variant B, i.e. after 2-stage ageing. In turn, extrapolation of results of creep tests to 10000 h conducted with graphical method showed that specimens of variant A, i.e. after 1-stage ageing showed higher creep resistance at the temperature of 650÷750°C.

The reason for lower creep resistance of the specimens heat treated according to variant B should be sought in a larger number of secondary phase particles precipitated on grain boundaries, which determines earlier initiation of the creep microcracks. This has been corroborated by the observation of the fractures morphology of specimens after creep where a faster development and larger fraction of intergranular cracks was found, indicating lower cohesion of the grain boundaries, especially at elevated temperature in the scope of 700÷750°C.

It was found that in creep conditions, failure development proceeded in a similar way in the microstructures of variant A and B specimens, and was concentrated mostly at grain boundaries and in boundary regions. At the first creep stage, creep pores nucleation occurred at grain boundaries and their coalescence. At the second stage, creep microcracks were formed at grain boundaries and first overageing effects were observed in the matrix. At the third creep stage, advanced intergranular cracking was detected at grain boundaries and in boundary regions, while distinct effects of overageing were found in the austenite matrix, connected with the $\gamma' \rightarrow \eta$ transition and formation of a number of transcrystalline lamellae of phase η in a Widmanstätten patter.

It was found that the substructure of specimens after creep of both variants of Fe-Ni alloy ageing depends basically on the creep temperature which determines also the creep mechanism. At the temperature of 650°C, creep is mainly of dislocation nature, and in the alloy substructure, increased density of dislocation was observed in a form of clusters, the movement of which was blocked by dispersive γ' phase particles. At higher temperature of 700°C, creep is of complex, dislocation-diffusive nature, and in the alloy substructure, both the strengthening effect and dynamic recovery effects were revealed as well as coagulation process of γ' phase particles. At the highest temperature of 750°C, creep is mainly of a diffusive nature, and in the alloy substructure, distinct effects of overageing are observed, connected with transition $\gamma' \rightarrow \eta$, accompanied with dissolution of the neighbouring γ' phase particles and a decrease of dislocation density in the matrix.

On the basis of the results obtained, a conclusion can be drawn that the studied Fe-Ni alloy is characterized by better material characteristics after solution heat treatment and 1-stage ageing at 715°C/16 h/air. With its slightly decreased strength properties, the alloy heat treated according to variant A shows definitely higher temporal creep strength, especially at the temperature within the range of 700÷750°C and extended operation time of ca. 10000 h.

The study shows a significant effect of the applied ageing variants on mechanical properties and fatigue life and creep resistance of the tested austenitic Fe-Ni superalloy. It was found that both, at the room and elevated temperatures, the specimens after 2-stage ageing were distinguished by higher strength properties. In a case of low-cycle fatigue tests carried out at a room and elevated temperatures and short creep tests at elevated temperatures, specimens after single-stage ageing were characterized with higher fatigue life and creep resistance. The obtained test results may be used to optimise heat treatment and forecast the operation conditions of products made out of Fe-Ni superalloy at room and an elevated temperature.

9. References

1. Sims Ch.T., Stoloff N.S., Hagel W.C.: The Superalloys II, A. Wiley Witescience Publications, New York 1987.

2. Stoloff N.S.: Wrought and P/M superalloys, In: ASM Handbook, Vol. 1: Properties and Selection Irons, Steels and High-Performance Alloys, ASM Materials Information Society, 1990, p. 950-977.

3. Morlet J.: Steels and Nickel-Base Alloys, In: High Temperature Alloys. Their Exploitable Potential, Elsevier Applied Science, London and New York 1985, p. 221-233.

4. Rohrbach K.P.: Trends in high-temperature alloys, Advanced Materials & Processes 10, 1995, p. 37-40.

5. Reed R.P., Tobler R.L., Mikesel R.P.: Fracture Toughness and Crack Growth Rate in Fe-Ni-Cr Alloy at 298, 76 and 4 K, In: Advances in Cryogenic Engineering, Plenum Press, New York and London 1983, p. 321-331.

6. Abe T., Kohno M., Suzuki A., Scanlan R.M.: Cryogenic Mechanical Properties of A286 Alloy and 304LN Stainless Steel Used in Fabrication of Support Struts for Superconducting Magnets, In: Steel forgings – Congresses, American Society for Testing and Materials, 1986.

7. Hiraga K., Ishikawa K., Tachikawa K., Yoshioka S., Inoue A., Takayanagi S.: Low Temperature Strength and Fatigue Properties of Welded Alloy A286, Trans. Iron Steel Inst. Jap. 26, 1986, B-256.

8. Harris D.R.: Physical Metallurgy of Fe-Cr-Ni Austenitic Steels, In: Mechanical and Nuclear Applications of Stainless Steel at Elevated Temperatures, The Metals Society, London, 1981, p. 9-33.

9. Kohno M., Yamada T., Ohta S., Aota K., Honjoh T.: Characteristics of a Heavy A286 Disk for Gas Turbines, Trans. Iron Steel Inst. Jap. 21, 1981, B-411.

10. Brooks J.W., Bridges P.J.: High Temperature Alloys for Gas Turbines and other Applications, D. Reidl Publish. Company, Netherlands, 1986.

11. Weßling W.: Festigkeitseigenschaften von Stählen für eine langzeitige Beanspruchung bei hohen Temperaturen, VDI Berichte, No. 600.4, 1987, s. 45-84.

12. Sczerzenie F., Maurer G.E.: Developments in Disc Materials, Materials Science and Technology 3, 1987, p. 733-742.

13. Franson I.A., Maurer J.R.: Applications of Fe-Ni-Cr-Mo-N Super-Austenitic Alloy Welded Tube and Pipe in Power Plant, Tube & Pipe Technology 2, 1989, p. 30-35.

14. Pitler R.K.: New Specialty Alloys: Advanced Materials for New Markets, Metallurgia Italiana 81,1989, p. 1041-1045.

15. Fujita A., Takeda Y., Fujikawa T., Yokota H., Hizume A., Honjo T., Okamura M.: Application of Modified A286 Iron Base Superalloy to USC Turbine Rotor, Proc. of 5th Liége Conference: Materials for Advanced Power Engineering, 1994, p. 515-526.

16. Tsuji H., Nakajima H., Kondo T.: Development of a New Ni-Cr-W Superalloy for Application to High-Temperature Structures, Proc. of 5th Liége Conference: Materials for Advanced Power Engineering, 1994, p. 939-948.

17. Härkegård G., Guédou J.Y.: Disc Materials for Advanced Gas Turbines, Proc. of the 6th Liége Conference: Materials for Advanced Power Engineering, 1998, p. 913-931.

18. Masuyama F.: Steam Plant Material Development in Japan, Proc. of the 6th Liége Conference: Materials for Advanced Power Engineering, 1998, p. 1807-1824.

19. Seth B.B.: Superalloys – the Utility Gas Turbine Perspective, Proc. of the Ninth International Symposium on Superaalloys: Superalloys 2000, A Publication of The Minerals, Metals & Materials Society, Pennsylvania, 2000, p. 3-16.

20. Ulianin E.A., Swistunowa T.W., Lewin F.L.: The corrosion resistant of iron- and nickel-base alloys, Mietallurgija, Moscow, 1986, p. 51-100 (in Russian).

21. Brooks J.A., Thompson A.W.: Microstructure and Hydrogen Effects on Fracture in the Alloy A-286, Metall. Trans. AIME 24A, 1993, p. 1983-1991.

22. Pound B.G.: The ingress of hydrogen into precipitation-hardened alloys A-286 and C17200, Corrosion Science 42, 2000, p. 1269-1281.

23. Mei Z., Morris J.W.: The Growth of Small Fatigue Cracks in A286 Steel, Metallurgical Transactions AIME 24A, 1993, p. 689-700.

24. Fricke W., Paul W., Hähner R., Eltner G., Müller H.: Beitrag zur Herstellung und Anwendung hochwarmfester Legierungen als Werkstoffe für Warmarbeitswerkzeuge, Neue Hütte 32, 1987, s. 55-59.

25. Kortmann W.A.: Extrusion tooling for N-F metals, Tube International, No. 9, 1989, p. 247-251.

26. Fricke W.: Verbesserung der Standzeit von Warmarbeitswerkzeugen durch den Einsatz aushärtbaren FeNiCr-, NiFeCr-, und NiCr-Legierungen, Stahlberatung 16, 1989, s. 10-13.

27. Pickering F.B.: Some aspects of the precipitation of nickel-aluminium-titanium intermetallic compounds in ferrous materials, Proc. Conference:

The Metallurgical Evolution of Stainless Steels, The Metals Society, London 1979, p. 391-401.

28. Schubert F.: Mechanische Eigenschaften von Superlegierungen und ihren Verbunden, VDI Berichte, No. 600.4, 1987, s. 85-136.

29. Stickler R.: Phase Stability in Superalloys, Proc. of the Symposium: High-Temperature Materials in Gas Turbines, Baden 1973, p. 115-146.

30. Pickering F.B.: Physical metallurgical development of stainless steels, Proc. of the Conference: Stainless Steels'84, Göteborg 1984, p. 1-28.

31. Maciejny A., Ducki K.J.: Changes in the structure and properties of the austenitic creep-resisting steels hardened precipitately by the intermetallic phases of the γ' type, Materials Engineering, Sigma NOT, Warsaw 3, 1990, p. 50-55 (in Polish).

32. Dulis E.J.: Age-hardening austenitic stainless steels, Proc. Conference: The Metallurgical Evolution of Stainless Steels, The Metals Society, London 1979, p. 420- 441.

33. Ducki K.: Precipitation of Intermetallic Phases in Cr-Ni-Ti-Al Austenitic Steel, Proc. IX Conference on Electron Microscopy of Solids, Kraków-Zakopane 1996, p. 491-494.

34. Ducki K., Hetmańczyk M.: The influence of prolonged ageing on the structure and properties of precipitation hardened austenitic alloy, Materials Engineering, Sigma NOT, Warsaw 4, 2001, p. 290-293.

35. Ducki K.J., Hetmańczyk M., Kuc D.: Analysis of precipitation process of the intermetallic phases in a high-temperature Fe-Ni austenitic alloy, Materials Chemistry and Physics 81, 2003, p. 490-492.

36. Ducki K.J.: Precipitation and growth of intermetallic phase in a high-temperature Fe-Ni alloy, Journal of Achievements in Materials and Manufacturing Engineering 18, 2006 p. 87-90.

37. Ducki K.J.: Structure and precipitation strengthening in a high-temperature Fe-Ni alloy, Archives of Materials Science and Engineering 28, 2007, p. 203-210.

38. Ducki K.J.: Analysis of the precipitation and growth processes in a high-temperature Fe-Ni alloy, Journal of Achievements in Materials and Manufacturing Engineering 31, 2008, p. 226-232.

39. Ducki K.J.: Microstructural aspects of deformation, precipitation and strengthening processes in austenitic Fe-Ni superalloy. Monograph. Copyright by Silesian University of Technology, 2010, p.1-136 (in Polish).

40. Schubert F.: Temperature and Time Dependent Transformation: Application to Heat Treatment of High Temperature Alloys, In: Phase Stability in High Temperature Alloys, Appied Science Publishers LTD, London 1981, p. 119-149.

41. Sourmail T.: Precipitation in creep resistant austenitic stainless steels, Materials Science and Technology 17, 2001, p. 1-14.

42. Thompson A.W., Brooks J.A.: The Mechanism of Precipitation Strengthening in an Iron-Base Superalloy, Acta Metallurgica 30, 1982, p. 2197-2203.

43. Larkin W.A., Lewin F.L., Rahsztadt A.G.: Influence of tungsten on properties of austenitic precipitation strengthened Fe-Cr-Ni alloys, Mietallowiedienije i termiczeskaja obrabotka mietallow 9, 1982, p. 50-52 (in Russian).

44. Watanabe R., Kuno T.: Alloy design of nickel-base precipitation hardened superalloys, Trans. Iron a. Steel Inst. Jap. 16, 1976, p. 437-446.

45. ASM Metals Reference Book: Heat- and Corrosion-Resistant Alloys, Copyright by the American Society for Metals, Metals Park, Ohio 1983, p. 399-418.

46. Ohta S., Igari S., Uchida H., Fujiwara M.: The Effect of Ti/Al Ratio on Creep Rupture Strength of Cold-Worked Gamma Prim Precipitation Hardened Fe-14Cr-30Ni Alloys for Fast Breeder Reactor Fuel Cladding Tubes, Trans. Iron. Steel Inst. Jap. 21, 1981, B.489.

47. Rho B.S., Nam S.W.: Fatigue-induced precipitates at grain boundary of Nb-A286 alloy in high temperature low cycle fatigue, Mater. Science & Engineer. A291, 2000, p. 54-59.

48. Kumar P.: Role of Niobium and Tantalum in Superalloys, In: Advances in High Temperature Structural Materials and Protective Coatings, Published by National Research Council of Canada, Ottawa 1994, p. 73-94.

49. Padilha A.F., Pohl M., Ramanathan L.V.: The Effect of Niobium Addition on the Microstucture of Fully Austenitic Fe-15%Cr-15%Ni Stainless Steels, Prakt. Metallogr. 31, 1994, p. 436-447.

50. Fujiwara M., Uchida H., Ohta S.: Effect of niobum content on creep strength of cold-worked 15Cr-15Ni-2.5Mo austenitic steel, Journal of Materials Science Letters 14, 1995, p. 297-301.

51. Quist W.E., Taggart R., Polonis D.H.: The Influence of Iron and Aluminium on the Precipitation of Metastable Ni_3Nb Phases in the Ni-Nb System, Metallurgical Transactions 2, 1971, p. 825-832.

52. Choudhury I.A., El-Baradie M.A.: Machinability of nickel-base super alloys: a general review, Journal of Materials Processing Technology 77, 1998, p. 278-284.

53. Loria E.A.: Recent Developments in the Progress of Superalloy 718, JOM 44, 1992, p. 33-36.

54. Inco Alloys International Inc.: Oxide Dispersion Strengthened Superalloys made by Mechanical Alloying, Publication No. IAI-9-2, 1993, p. 1-8.

55. Schröder J.H., Arzt E.: Electron-Microscopic Investigations of Dispersion-Strengthened Superalloys, Prakt. Metallogr. 25 (6), 1988, p. 264-273.

56. Wasilkowska A., Bartsch M., Messerschmidt U., Herzog R., Czyrska-Filemonowicz A.: Creep mechanisms of ferritic oxide dispersion strengthened alloys, Journal of Materials Processing Technology 133, 2003, p. 218-224.

57. McColvin G., Munasinghe D., O'Driscoll J., Jacobs M.: Fabrication of gas turbine combustion hardware in ODS ferritic materials, Proc. of the 7th Liege Conference: Materials for Advanced Power Engineering 21, 2002, p. 833-844.

58. Pérez P.: Influence of the alloy grain size on the oxidation behaviour of PM2000 alloy, Corrosion Science 44 (8), 2002, p. 1793-1808.

59. Capdevila C., Chen Y.L., Jones A.R., Bhadeshia K.D.H.: Grain Boundary Mobility in Fe-Base Oxide Dispersion Strengthened PM2000 Alloy, ISIJ International 43 (5), 2003, p. 777-783.

60. Suryanarayana C.: Mechanical alloying and milling, Progress in Materials Science 46, 2001, p. 1-184.

61. Brandis H., Huchtemann B.: Technologie der Wärmebehandlung warmfeste und hochwarmfester Stähle, Thyssen Edelst. Techn. Ber. 1, 1981, s. 28-40.

62. Ducki K.J., Cieśla M.: Effect of heat treatment on the structure and fatigue behaviour of austenitic Fe-Ni alloy, Journal of Achievements in Materials and Manufacturing Engineering 30, 2008, p. 19-26.

63. Okrajni J., Cieśla M., Swadźba L.: High-temperature low-cycle fatigue and creep behaviour of nickel-based superalloys with heat-resistant coating, Fatigue and Fracture of Materials and Engineering Structures 21, 1998, p. 947-954.

64. Gronostajski Z., Jaśkiewicz K.: Influence of monotonic and cyclic deformation sequence on behavior of CuSi3.5 silicon bronze, Journal of Achievements in Materials and Manufacturing Engineering 15, 2006, p. 39-46.

65. Okrajni J., Marek A., Junak G.: Description of the deformation process under thermo-mechanical fatigue, Journal of Achievements in Materials and Manufacturing Engineering 21, 2007, p. 15-23.

66. Okrajni J., Marek A., Junak G.: Stress-strain characteristics under mechanical and thermal loading, Journal of Achievements in Materials and Manufacturing Engineering 20, 2007, p. 271-274.

67. Ducki K.J.: Effect of heat treatment on the structure and creep resistance of austenitic Fe-Ni alloy, Archives of Materials Science and Engineering 47, 2011, p. 33-40.

68. Kocańda S.: A fatigue cracking of metals, WNT, Warsaw, 1985 (in Polish).

69. Hernas A.: Creep Strength of Steels and Alloys, Publishing by Silesian University of Technology, Gliwice, 2000 (in Polish).

70. Penkalla H.J., Schubert F.: Ni-base wrought alloy development for USC steam turbine rotorapplications, Materials Engineering, Sigma NOT, Warsaw 3 (2004) 415-421.

Printed by Books on Demand GmbH, Norderstedt / Germany